D4d REC

WITHDRAWN

Receptor Binding Techniques

METHODS IN MOLECULAR BIOLOGY™

John M. Walker, SERIES EDITOR

316. **Bioinformatics and Drug Discovery**, edited by *Richard S. Larson, 2005*
315. **Mast Cells:** *Methods and Protocols,* edited by *Guha Krishnaswamy and David S. Chi, 2005*
314. **DNA Repair Protocols:** *Mammalian Systems, Second Edition,* edited by *Daryl S. Henderson, 2005*
313. **Yeast Protocols:** *Second Edition,* edited by *Wei Xiao, 2005*
312. **Calcium Signaling Protocols:** *Second Edition,* edited by *David G. Lambert, 2005*
311. **Pharmacogenomics:** *Methods and Applications,* edited by *Federico Innocenti, 2005*
310. **Chemical Genomics:** *Reviews and Protocols,* edited by *Edward D. Zanders, 2005*
309. **RNA Silencing:** *Methods and Protocols,* edited by *Gordon Carmichael, 2005*
308. **Therapeutic Proteins:** *Methods and Protocols,* edited by *C. Mark Smales and David C. James, 2005*
307. **Phosphodiesterase Methods and Protocols,** edited by *Claire Lugnier, 2005*
306. **Receptor Binding Techniques:** *Second Edition,* edited by *Anthony P. Davenport, 2005*
305. **Protein–Ligand Interactions:** *Methods and Applications,* edited by *G. Ulrich Nienhaus, 2005*
304. **Human Retrovirus Protocols:** *Virology and Molecular Biology,* edited by *Tuofu Zhu, 2005*
303. **NanoBiotechnology Protocols,** edited by *Sandra J. Rosenthal and David W. Wright, 2005*
302. **Handbook of ELISPOT: Methods and Protocols,** edited by *Alexander E. Kalyuzhny, 2005*
301. **Ubiquitin–Proteasome Protocols,** edited by *Cam Patterson and Douglas M. Cyr, 2005*
300. **Protein Nanotechnology:** *Protocols, Instrumentation, and Applications,* edited by *Tuan Vo-Dinh, 2005*
299. **Amyloid Proteins:** *Methods and Protocols,* edited by *Einar M. Sigurdsson, 2005*
298. **Peptide Synthesis and Application,** edited by *John Howl, 2005*
297. **Forensic DNA Typing Protocols,** edited by *Angel Carracedo, 2005*
296. **Cell Cycle Control:** *Mechanisms and Protocols,* edited by *Tim Humphrey and Gavin Brooks, 2005*
295. **Immunochemical Protocols,** *Third Edition,* edited by *Robert Burns, 2005*
294. **Cell Migration:** *Developmental Methods and Protocols,* edited by *Jun-Lin Guan, 2005*
293. **Laser Capture Microdissection:** *Methods and Protocols,* edited by *Graeme I. Murray and Stephanie Curran, 2005*
292. **DNA Viruses:** *Methods and Protocols,* edited by *Paul M. Lieberman, 2005*
291. **Molecular Toxicology Protocols**, edited by *Phouthone Keohavong and Stephen G. Grant, 2005*
290. **Basic Cell Culture Protocols**, *Third Edition,* edited by *Cheryl D. Helgason and Cindy L. Miller, 2005*
289. **Epidermal Cells**, *Methods and Applications,* edited by *Kursad Turksen, 2005*
288. **Oligonucleotide Synthesis**, *Methods and Applications,* edited by *Piet Herdewijn, 2005*
287. **Epigenetics Protocols**, edited by *Trygve O. Tollefsbol, 2004*
286. **Transgenic Plants:** *Methods and Protocols,* edited by *Leandro Peña, 2005*
285. **Cell Cycle Control and Dysregulation Protocols:** *Cyclins, Cyclin-Dependent Kinases, and Other Factors,* edited by *Antonio Giordano and Gaetano Romano, 2004*
284. **Signal Transduction Protocols,** *Second Edition,* edited by *Robert C. Dickson and Michael D. Mendenhall, 2004*
283. **Bioconjugation Protocols,** edited by *Christof M. Niemeyer, 2004*
282. **Apoptosis Methods and Protocols,** edited by *Hugh J. M. Brady, 2004*
281. **Checkpoint Controls and Cancer, Volume 2:** *Activation and Regulation Protocols,* edited by *Axel H. Schönthal, 2004*
280. **Checkpoint Controls and Cancer, Volume 1:** *Reviews and Model Systems,* edited by *Axel H. Schönthal, 2004*
279. **Nitric Oxide Protocols,** *Second Edition,* edited by *Aviv Hassid, 2004*
278. **Protein NMR Techniques,** *Second Edition,* edited by *A. Kristina Downing, 2004*
277. **Trinucleotide Repeat Protocols,** edited by *Yoshinori Kohwi, 2004*
276. **Capillary Electrophoresis of Proteins and Peptides,** edited by *Mark A. Strege and Avinash L. Lagu, 2004*
275. **Chemoinformatics,** edited by *Jürgen Bajorath, 2004*
274. **Photosynthesis Research Protocols,** edited by *Robert Carpentier, 2004*
273. **Platelets and Megakaryocytes, Volume 2:** *Perspectives and Techniques,* edited by *Jonathan M. Gibbins and Martyn P. Mahaut-Smith, 2004*
272. **Platelets and Megakaryocytes, Volume 1:** *Functional Assays,* edited by *Jonathan M. Gibbins and Martyn P. Mahaut-Smith, 2004*
271. **B Cell Protocols,** edited by *Hua Gu and Klaus Rajewsky, 2004*
270. **Parasite Genomics Protocols,** edited by *Sara E. Melville, 2004*
269. **Vaccina Virus and Poxvirology:** *Methods and Protocols,*edited by *Stuart N. Isaacs, 2004*
268. **Public Health Microbiology:** *Methods and Protocols,* edited by *John F. T. Spencer and Alicia L. Ragout de Spencer, 2004*

METHODS IN MOLECULAR BIOLOGY™

Receptor Binding Techniques

Second Edition

Edited by

Anthony P. Davenport

Clinical Pharmacology Unit, University of Cambridge, Centre for Clinical Investigation, Addenbrooke's Hospital, Cambridge, UK

HUMANA PRESS ✷ TOTOWA, NEW JERSEY

© 2005 Humana Press Inc.
999 Riverview Drive, Suite 208
Totowa, New Jersey 07512

www.humanapress.com

All rights reserved. No part of this book may be reproduced, stored in a retrieval system, or transmitted in any form or by any means, electronic, mechanical, photocopying, microfilming, recording, or otherwise without written permission from the Publisher. Methods in Molecular Biology™ is a trademark of The Humana Press Inc.

All papers, comments, opinions, conclusions, or recommendations are those of the author(s), and do not necessarily reflect the views of the publisher.

This publication is printed on acid-free paper. ∞

ANSI Z39.48-1984 (American Standards Institute)

Permanence of Paper for Printed Library Materials.

Production Editor: Nicole E. Furia

Cover design by Patricia F. Cleary

Cover Illustration by Peter Johnström and Anthony Davenport: Imaging of endothelin (ET) receptor distribution in rat kidney following *in vivo* infusion of [18F]-ET-1. Binding of [18F]-ET-1 to ET receptors in substructures of rat kidney (shown on the right, see Chapter 11) visualised *in vivo* using positron emission tomography and subsequently *ex vivo* following tissue sectioning and phosphor imaging autoradiography, shown on the left (see Chapter 10) .

For additional copies, pricing for bulk purchases, and/or information about other Humana titles, contact Humana at the above address or at any of the following numbers: Tel.: 973-256-1699; Fax: 973-256-8341; E-mail: orders@humanapr.com; or visit our Website: www.humanapress.com

Photocopy Authorization Policy:
Authorization to photocopy items for internal or personal use, or the internal or personal use of specific clients, is granted by Humana Press Inc., provided that the base fee of US $30.00 per copy is paid directly to the Copyright Clearance Center at 222 Rosewood Drive, Danvers, MA 01923. For those organizations that have been granted a photocopy license from the CCC, a separate system of payment has been arranged and is acceptable to Humana Press Inc. The fee code for users of the Transactional Reporting Service is: [1-58829-420-X/05 $30.00].

Printed in the United States of America. 10 9 8 7 6 5 4 3 2 1

eISBN 1-59259-927-3

ISSN 1064-3745

Library of Congress Cataloging-in-Publication Data

Receptor binding techniques / edited by Anthony P. Davenport.— 2nd ed.
p. cm. — (Methods in molecular biology)
Includes bibliographical references and index.
ISBN 1-58829-420-X (alk. paper)
1. Receptor-ligand complexes—Research—Laboratory manuals. 2. Radioligand assay—Laboratory manuals. 3. Cell receptors—Research—Laboratory manuals. I. Davenport, Anthony P. II. Series.
QH603.C43R39 2005
571.6—dc22
2004025577

UNIVERSITY OF BRISTOL
LIBRARY
MEDICAL

Preface

The objective of *Receptor Binding Techniques, Second Edition* is to provide detailed practical methods for studying of receptors *in silico*, in vitro, and in vivo. The sequencing of the human genome has largely been completed. In addition to the established families of receptors, more than one hundred gene sequences have been predicted to encode novel G protein-coupled receptors (excluding olfactory receptors) together with a much smaller number of tyrosine kinase and nuclear receptors. Initially these are designated "orphan" receptors since their activating ligand has not been identified. Many sequences encoding orphan receptors have been artificially expressed in cell lines and are being paired with their cognate endogenous ligands by screening compounds from existing libraries or from tissue extracts. An unprecedented number of new receptor systems are emerging to be explored through ligand binding techniques. In addition, there is an expanding wealth of animal models such as disruption of genes in mice (knock-outs or knock-ins) in which to apply these techniques as a means to unravel the role of established or emerging orphan receptors.

In addition to curated databases of sequences encoding receptors, a large body of experimental data on ligand receptor interactions is available from public websites. These are dedicated to the major families together with sites concerned with downstream processes in intracellular, second messenger signaling. The efficient mining of these databases described in the first chapter is a logical first step in researching novel and established receptor systems. Chapter 2 provides detailed methodological information for the pairing of "orphan" receptors to identify their cognate ligand by fluorometric imaging plate reader assays that are amenable to high throughput screening. Molecular techniques can provide unequivocal evidence for the presence of mRNA encoding a specific receptor in a particular tissue. The quantitative measurement of mRNA in homogenates of tissue or cells using real-time polymerase chain reaction assays (Chapter 3) and determining the precise cellular or anatomical localization by *in situ* hybridization in tissue sections or Northern analysis in homogenates (Chapter 4) can provide initial clues for a role of de-orphanized receptors recently paired with their cognate ligands, narrowing the search for function.

The three principal radioligand binding assays used to characterize receptors (saturation, competition, and kinetic) are described in Chapter 5. Analysis by iterative curve-fitting programs provides key binding parameters: the equilibrium dissociation constant, receptor density, and Hill slope. A powerful alternative to measure these parameters is the use of scintillation proximity assays (Chapter 6), which avoids the need to separate bound from free radioligand, particularly when rapid, high-throughput screening is required. Chapter 7 widens the scope of radiolabeled binding to encompass studying the distribution of enzymes, allosteric modulators of ion channels, and second messengers by either macro autoradiography or higher resolution micro autoradiography. A key protocol is provided for measuring unlabeled agonist-enhanced binding of [^{35}S]-GTPγS, widely used in the mapping of G protein-coupled receptors that can also be exploited when radiolabeled analogs have not yet been developed for a particular transmitter molecule.

Immunocytochemistry (Chapter 8) exploiting selective antisera to map the distribution of receptor protein, compliments radioligand binding, particularly in permitting the precise identification of cell types expressing a specific receptor by dual labeling visualized by confocal microscopy. Applications include the characterization of mice following targeted disruption of gene-encoding receptors. The analysis of ligand receptor interactions in living cells at the subcellular level by confocal microscopy is the focus of Chapter 9. Protocols are given for visualizing intracellular trafficking in real time with fluorescent-labeled ligands or with the receptor itself labeled with green fluorescent protein. This chapter also describes techniques for measuring dimerization of receptors or interaction with other proteins by fluorescence resonance energy transfer.

Positron emission tomography (PET) is the only quantitative technique with sufficient sensitivity to detect the binding of radiolabeled ligand to receptors in living animals. The most widely used positron-emitting isotopes used to label ligands, ^{11}C and ^{18}F, have short half-lives and are difficult to detect by conventional film-based autoradiography. The penultimate chapter (10) outlines strategies for measuring binding parameters in vitro by phosphor imaging to evaluate the sensitivity and specificity of novel PET ligands for use in vivo. Following the development of tomographs specifically designed for rodents that can achieve remarkably high resolution, the final chapter (11) considers applying these ligands to functional imaging of receptors in vivo.

I am very grateful to the authors for their excellent contributions to this second edition of *Receptor Binding Techniques*, reflecting the success of the first edition edited by Mary Keen.

Anthony P. Davenport

Contents

Contributors

ZHAOHUI AO • *Department of Vascular Biology and Thrombosis, Cardiovascular and Urogenital Center of Excellence for Drug Discovery, GlaxoSmithKline, King of Prussia, PA*

NICHOLAS M. BARNES • *Cellular and Molecular Neuropharmacology Research Group, Division of Neuroscience, Department of Pharmacology, The Medical School, University of Birmingham, Edgbaston, Birmingham, UK*

OLIVIER BARRET • *Wolfson Brain Imaging Centre, University of Cambridge, Addenbrooke's Hospital, Cambridge, UK*

JENNY BERRY • *The Maynard Centre, GE Healthcare Biosciences, Cardiff, UK*

MICHAEL BEYERMANN • *Forschungsinstitut für Molekulare Pharmakologie, Berlin, Germany*

JOHN P. COGSWELL • *Quantitative Expression, Genomic and Proteomic Sciences, Genetics Research, GlaxoSmithKline, Research Triangle Park, NC*

ANTHONY P. DAVENPORT • *Clinical Pharmacology Unit, University of Cambridge, Centre for Clinical Investigation, Addenbrooke's Hospital, Cambridge, UK*

STEPHEN A. DOUGLAS • *Department of Vascular Biology and Thrombosis, Cardiovascular and Urogenital Center of Excellence for Drug Discovery, GlaxoSmithKline, King of Prussia, PA*

TERRI J. DOVER • *Cellular and Molecular Neuropharmacology Research Group, Division of Neuroscience, Department of Pharmacology, The Medical School, University of Birmingham, Birmingham, UK*

MARK D. FIDOCK • *Target Genomics, Pfizer Global Research and Development, Sandwich, Kent, UK*

TIM D. FRYER • *Wolfson Brain Imaging Centre, University of Cambridge, Addenbrooke's Hospital, Cambridge, UK*

PAUL A. INSEL • *Department of Pharmacology, University of California San Diego, La Jolla, CA*

DOUGLAS G. JOHNS • *Department of Vascular Biology and Thrombosis, Cardiovascular and Urogenital Center of Excellence for Drug Discovery, GlaxoSmithKline, King of Prussia, PA*

PETER JOHNSTRÖM • *Clinical Pharmacology Unit, University of Cambridge, Centre for Clinical Investigation, Addenbrooke's Hospital, Cambridge, UK*

RHODA E. KUC • *Clinical Pharmacology Unit, University of Cambridge, Centre for Clinical Investigation, Addenbrooke's Hospital, Cambridge, UK*

Kristeen Maniscalco • *Department of Investigative Biology, Cardiovascular and Urogenital Center of Excellence for Drug Discovery, GlaxoSmithKline, King of Prussia, PA*

Paul Murdock • *Quantitative Expression, Genomic and Proteomic Sciences, Genetics Research, GlaxoSmithKline, Stevenage, Herts, UK*

Alexander Oksche • *Institut für Pharmakologie, Charité Universitätsmedizin Berlin, Thielallee, Berlin, Germany*

Rachel M. C. Parker • *The British Heart Foundation, London, UK*

Lisa Patel • *Department of Atherosclerosis, Cardiovascular and Urogenital Center of Excellence for Drug Discovery, GlaxoSmithKline, Stevenage, Herts, UK*

Molly Price-Jones • *The Maynard Centre, GE Healthcare Biosciences, Cardiff, UK*

Alessandra P. Princivalle • *Cellular and Molecular Neuropharmacology Research Group, Division of Neuroscience, Department of Pharmacology, The Medical School, University of Birmingham, Birmingham, UK*

Brinda K. Rana • *Department of Psychiatry, University of California San Diego, La Jolla, CA*

Hugh K. Richards • *Academic Neurosurgery Unit, University of Cambridge, Addenbrooke's Hospital, Cambridge, UK*

Nicola M. Robas • *Target Genomics, Pfizer Global Research and Development, Sandwich, Kent, UK*

Lea Sarov-Blat • *Department of Bioinformatics, GlaxoSmithKline, King of Prussia, PA*

Sheila Seepersaud • *Quantitative Expression, Genomic and Proteomic Sciences, Genetics Research, GlaxoSmithKline, Collegeville, PA*

Klaudia M. Steplewski • *Quantitative Expression, Genomic and Proteomic Sciences, Genetics Research, GlaxoSmithKline, Collegeville, PA*

David A. Walsh • *Academic Rheumatology, University of Nottingham, City Hospital, Nottingham, UK*

John Wharton • *Section on Experimental Medicine and Toxicology, Imperial College London, London, UK*

Burkhard Wiesner • *Forschungsinstitut für Molekulare Pharmakologie, Berlin, Germany*

Robert N. Willette • *Department of Investigative Biology, Cardiovascular and Urogenital Center of Excellence for Drug Discovery, GlaxoSmithKline, King of Prussia, PA*

1

Receptor Databases and Computational Websites for Ligand Binding

Brinda K. Rana and Paul A. Insel

1. Introduction

Ligand binding to receptors is a key step in the regulation of cellular function by neurotransmitters, hormones, and many drugs. Accordingly, this initial event in ligand action is important for understanding disease and designing new drugs. A large body of experimental data describing receptor–ligand interactions exists and is derived from studies of native and transfected cell systems, including a growing number of studies with artificial or naturally occurring receptor mutants. Taken together, genes encoding various receptors appear to form the largest classes of functional genes in mammalian genomes. This large number of genes and gene products, together with the expanding pool of ligands, provides, and will generate in the future, a huge amount of data. Such compilations of data create a need for comprehensive, web-based resources that compile and integrate information on receptor protein and nucleotide sequences, classification, experimental results, and computational tools for modeling interactions. A number of websites in the public domain provide useful data-mining tools and contain information on specific families of receptors or receptor subfamilies, such as the G protein-coupled receptors (GPCRs) (*1,2*), nuclear receptors, ion channel receptors, and others. A number of websites provide tools by which potential functions and molecular interactions can be derived to guide experimentalists in studies of receptor–ligand interaction and thus aid in defining the function of the receptor of interest. The goal of this chapter is to identify websites containing information that can facilitate

From: *Methods in Molecular Biology, vol. 306: Receptor Binding Techniques: Second Edition*
Edited by: A. P. Davenport © Humana Press Inc., Totowa, NJ

both computational and experimental studies of receptor–ligand interactions. We will identify and briefly review websites and certain software that are available for several different classes of receptors and their ligands.

2. Computational Websites and Software for Predicting Receptor–Ligand Binding

The search for new drugs, ligands of orphan receptors, and targets of ligands or drugs can be enhanced by computational tools. Researchers interested in drug design and receptor–ligand interactions can take advantage of the many computational resources facilitated by the rapidly expanding pool of structural data in the Protein Data Bank (PDB; http://www.rcsb.org/pdb) (*3*). At the time of preparation of this chapter (Fall, 2004), the three-dimensional (3-D) structure database of PDB contained 28,903 structures. The recent growth of the PDB and other databases of protein and nucleotide sequences have enhanced the development of computational tools for receptor function and receptor–ligand interactions. For example, a widely used structure-based drug design tool is the docking/scoring programs that predict putative ligands for a receptor of interest from large databases of molecules (*4,5*). Also, algorithms that consider computational sequence and structural comparisons of an uncharacterized receptor with previously characterized receptors can be used to suggest experiments required to define ligand-receptor interaction of an orphan receptor. Further, with the recent influx of human genetic variation data in pharmacogenomic studies, as well as data from mutational analyses, knowledge of amino acid residues important in ligand–receptor binding is becoming increasingly valuable in drug design; the resources described here can be used to predict the putative functional role of such variants. **Table 1** lists freely available resources of this type.

Relibase (http://Relibase.rutgers.edu or http://Relibase.ebi.ac.uk) is a web-based tool designed to facilitate data mining for protein–ligand related interaction (*6*). In particular, it enables the search and analysis of 3-D protein–ligand complexes in the PDB. Functions of Relibase include: the detailed analysis of superimposed ligand-binding sites; ligand similarity and substructure searches; and 3-D searches for protein–ligand and protein–protein interaction patterns. Relibase provides a resource for many classes of receptors and their ligands.

When the 3-D structure is available, computer-aided structure-based drug design ligand–protein docking/scoring programs are also useful. This method involves the docking of molecules in multiple conformations into receptor- binding sites to find potential ligands. Selection is based on a molecular binding score that evaluates some form of interaction energy between the docked molecule and the receptor. Many approaches exist and are being developed to improve scoring. The Ligand–Protein DataBase (LPDB; http://lpdb.scripps.edu (*7*)

Table 1
Computational Websites and Tools

Tool or Website	URL
LigBase	http://alto.compbio.ucsf.edu/ligbase/
Relibase	http://Relibase.rutgers.edu http://Relibase.ebi.ac.uk
CLiBE	http://xin.cz3.nus.edu.sg/group/CLiBE/CLiBE.asp
Protein Ligand Database	http://www-mitchell.ch.cam.ac.uk/pld/index.html
LIGPLOT	http://www.biochem.ucl.ac.uk/bsm/ligplot/ligplot.html
AutoDock	http://www.scripps.edu/pub/olson-web/doc/autodock/
DOCK	http://dock.compbio.ucsf.edu/

URL, uniform resource locator; CLiBE, Computed Ligand Binding Energy Database.

is designed to allow the selection of complexes based on various properties of receptors and ligands in order to assess or improve on existing scoring functions or to develop new scoring functions.

For those interested in docking/scoring programs that identify putative ligands or receptors of interest, a widely used docking program is DOCK (http://dock.compbio.ucsf.edu/) (*8*), which can be used to search databases of molecular structures for compounds which bind to particular receptors. AutoDock (http://www.scripps.edu/pub/olson-web/doc/autodock/) (*9*) is a suite of automated docking tools. It is designed to predict how small molecules, such as substrates or drug candidates, bind to a receptor of known 3-D structure. The Computed Ligand Binding Energy Database (CLiBE; http://xin.cz3.nus.edu.sg/group/CLiBE/CLiBE.asp) ***(10)*** is a freely accessible resource on the web. Because competition with natural ligands can affect drug binding, CLiBE uses a scoring system that accounts for such competitive interactions by using ligand-bound 3-D structures in the PDB and contains information about ligand function, properties, and computed energy. This database has been used to assess drug-resistant mutations in proteins, and in the analysis of binding competition in the prediction of therapeutic and toxicity targets of drugs (***11,12***). CLiBE contains 2803 distinctive ligand entries and 2256 distinctive receptor entries.

Sequence-based similarity searches provide a method for detecting functional sites of proteins. For example, PROSITE (http://us.expasy.org/prosite/) is a helpful tool in finding common binding or active sites or receptors. Understanding the importance of a single residue is facilitated by the consideration of the residue's location in 3-D space with respect to the ligand because the structural properties of a binding site can be conserved although the sequence diverges. LigBase (http://alto.compbio.ucsf.edu/ligbase/) (*13*) is a resource that combines sequence and structural information in a database format. It can be used to compare known and potential binding sites in experimentally determined protein structures, and provides a resource for the analysis of families of related binding sites. This database contains approx 50,000 ligand-binding sites for small molecules found in the PDB, and it summarizes ligand data with structural information from PDB and graphically depicts residues in binding sites for comparison with other structurally defined family members. LigBase utilizes another useful program, LIGPLOT (available in the public domain at http://www.biochem.ucl.ac.uk/bsm/ligplot/ligplot.html) (*14*), which generates 2-D schematic diagrams of protein-ligand interactions from the 3-D coordinates of a given PDB file in order to generate diagrams of binding sites. Another algorithm based on protein sequence comparisons is the Evolutionary Trace (http://imgen.bcm.tmc.edu/molgen/labs/lichtarge/lab.html) (*15*) which, in order to decipher interactions between proteins and to identify specific drug targets, predicts functional sites that mediate protein binding. These predictions are based on the relative functional importance of amino acid residues in a protein sequence by correlating variations during evolution with divergences in the phylogenetic tree of the family in which the protein resides. This approach exploits the natural mutational scanning that occurred in evolutionary history.

3. Databases Dedicated to Specific Receptors

In this section, we discuss databases for several important classes of receptors with an emphasis on GPCRs and nuclear receptors.

3.1. G Protein-Coupled Receptors

As the largest receptor family in the human genome and the target of a large percentage of currently used drugs, GPCRs have attracted considerable interest. There are upward of 1000 GPCR genes in various genomes; these include receptors for neurotransmitters, hormones, light, odorants, and tastes. Whereas some of these genes encode receptors for known ligands, a substantial number have been identified as "orphans" because of the absence of clearly identifiable (natural) agonist ligands. A list of GPCRs from various species with links to protein sequence (2148 total entries and 665 human entries at time of writing) can be found at ExPASy (http://www.expasy.org/cgi-bin/lists?7tmrlist.txt). **Table 2** provides a listing of a number of websites and databases with information regarding GPCRs. Although there are many more GPCR-specific databases than listed here, we limit this to available databases and tools for the study of GPCR ligand binding and classification. For a detailed list of GPCR websites, *see* Rana and Insel, 2002 (*2*).

Table 2
G Protein-Coupled Receptor (GPCR) Specific Websites

Tool or Website	URL
GPCR database	http://www.gpcr.org/
GRAP	http://tinygrap.uit.no/GRAP/homepage.html
GPCR Pattern Recognition	http://www.biochem.ucl.ac.uk/bsm/dbbrowser/GPCR/
Viseur Program	http://transport.physbio.mssm.edu/viseur/viseur.html
Olfactory Receptor Database	http://senselab.med.yale.edu/senselab/ORDB/default.asp
Cytokine Signaling Pathway Database	http://cytokine.medic.kumamoto-u.ac.jp/

URL, uniform resource locator.

The International Union of Pharmacology maintains the IUPHAR Receptor Database (http://iuphar-db.org/iuphar-rd/index.html). This is a key on-line curated repository of data characterizing human, rat, and mouse receptors as well as definitive information on receptor classification. It includes ligand information and agonist and antagonist potencies for a variety of GPCRs, links to more detailed compendia and papers available online, and can be a good starting point for studying GPCRs.

The GPCR database (GPCRDB) (http://www.gpcr.org/) (***16***) is another good starting point for studying GPCRs and ligand interactions. The most comprehensive GPCR database tool available in the public domain, GPCRDB integrates sequence data, evolutionary relationships, mutation information, and pharmacological data. GPCRDB contains a "Ligand Dissociation Constants for GPCRs" database, which incorporates data curated by two independent sources. Browsing through a list of 300 ligands, one can obtain "target receptor" information along with dissociation constants for the ligand–receptor interaction. The GPCRDB also contains atomic coordinates of 3-D models for GPCRs; such data can be useful in working with the computational tools described in **Subheading 1**. For those interested in the study of genetic variants of GPCRs, an updated list of 7080 point mutations extracted from the scientific literature is available in GPCRDB.

The GRAP mutant databases (http://tinygrap.uit.no/GRAP/homepage.html) (***17***) consist of the oGRAP and the tinyGRAP. TinyGRAP contains information on about 10,500 mutants from approx 1380 scientific publications. A query of a GPCR yields information on binding and other assays with reference links for published mutagenesis studies and genetic variants. The oGRAP, an older version to tinyGRAP, contains lists of constitutively active mutants, lists of publications organized according to GPCR class, and a list of agonist and antagonist ligands that are classified according to receptor class. Quantitative data (K_i or IC_{50}) for competitive ligands and K_d values for radioligands are also available.

For structural modeling, Swiss-Model has a GPCR mode (http://www.expasy.org/swissmod/SWISS-MODEL.html) and is a fully automated protein structure homology-modeling server. For managing and visualizing GPCRs, the Viseur program (http://transport.physbio.mssm.edu/viseur/viseur.html) is accessible on-line and enables 3-D and Snakelike Plots, the latter to accommodate the 7-α-helical, membrane-spanning topology of GPCRs. On the basis of an alignment and a template protein (PDB file), one can build a model and view and transform it (rotate, translate, or move helices in the model). Each residue of the model can be linked to information, such as mutant data from tinyGRAP or user annotation. This unique feature, combined with the raw model creation (from the alignment editor), enables the rapid construction of a variety of schematic models.

Classification of the GPCR of interest can help investigators in defining molecules that will interact with particular receptors. Computational probes of the human genome for previously unreported GPCRs have identified a large number of orphan GPCRs (***18***). The most current classification of GPCRs in the human genome has been reported by Fredrikkson et al. (***19***). Characterizing a GPCR involves a combination of identifying both its ligand and the coupled G protein, and can be complicated by GPCRs that bind to the same ligand but couple to different G proteins, or GPCRs coupling to the same G protein binding different ligands. Several computational tools have been developed that use different algorithms to classify uncharacterized GPCRs based on ligand binding and other properties. The GPCR pattern recognition (http://www.biochem.ucl.ac.uk/bsm/dbbrowser/GPCR/) is a diagnostic resource that enables the user to search a query sequence against a fingerprint database to determine if it belongs to a particular GPCR superfamily, family, or receptor subtype (approx 120 fingerprints are available). This approach can be used to help predict the class of a particular orphan GPCR and to help narrow the type or range of potential ligands to be tested. GPCR Coupling Specificity Website (http://ep.ebi.ac.uk/GPCR/) (***20***) aids the investigator in predicting which effector the receptor will couple to following receptor activation through ligand binding. Another method uses a phylogenetic tree based profile hidden Markov model to classify GPCRs by ligand and G protein (http://mathbio.nimr.mrc.ac.uk/goldstein/GPCR/) (***21***).

There are also several websites dedicated to GPCR subfamilies and their ligands. One that is especially useful for the study of ligand–receptor binding is the Olfactory Receptor Database (ORDB; http://senselab.med.yale.edu/

senselab/ORDB/default.asp) (*22*) ORDB provides tools for investigators to analyze the functions of the very large olfactory family of GPCRs. A list of ligands can be browsed through ORDB or the sister-site odorDB (http://senselab.med.yale.edu/senselab/odordb/) to obtain chemical information on the molecules and, via links, information on their target receptors. The Cytokine Signaling Pathway Database (http://cytokine.medic.kumamoto-u.ac.jp/) provides biochemical data and references regarding signaling molecules and ligand–receptor relationships for cytokines and their receptors.

3.2. Nuclear Receptors

Several resources exist for studying nuclear receptors, binding, and structure (**Table 3**). These include three that we will discuss here: (1) The "Nuclear Receptor Resource"; (2) The Nuclear Receptor Database; and (3) the Nuclear Receptor Mutation Database.

The Nuclear Receptor Resource (NRR; http://nrr.georgetown.edu/NRR/nrrhome.htm) (*23*) is designed to disseminate information on new techniques, new vectors, and other technical observations related to nuclear receptors. An additional, potentially quite useful category is the inclusion of descriptions of negative results that would not generally be reported in conventional journal articles but which could be of assistance to workers in the field. Currently contained within the NRR are resources for seven receptor subfamilies:

- The Androgen Receptor Gene Mutations Database WWW Server (http://www.androgendb/.mcgill.ca)
- The Estrogen Receptor Resource (ERR; http://nrr.georgetown.edu/Estrogen%20Receptor/ER-PAGE/Main.html)
- The Glucocorticoid Receptor Resource (GRR; http://nrr.georgetown.edu/GRR/grr1.htm)
- The Peroxisome Proliferator-Activated Receptor (PPAR) Resource (http://www.cas.psu.edu/docs/CASDEPT/VET/jackvh/ppar/pparrfront.htm)
- The Steroid Receptor Resource(http://nrr.georgetown.edu/NRR/srapr/srapr.html)
- The Thyroid Receptor Resource (http://nrrgeorgetown.edu/NRR/TRR/trrfront.html),
- The Vitamin D Receptor Resource (http://vdr.bu.edu)

A unique feature of the Androgen Receptor Gene Mutations Database WWW Server is the inclusion of a database of naturally occurring human androgen receptor variants and somatic mutations with associated phenotypes and experimentally derived binding properties compiled from the literature. Also featured is a 3-D model for the androgen receptor and maps and tables of androgen receptor interacting proteins. The ERR provides a variety of descriptive information on estrogen receptors as well as sequence alignments and knock-out phenotypes. Similarly to these two databases, the GRR, PPAR

Table 3
Websites for Nuclear Receptors

Tool or Website	URL
Nuclear Receptor Resource	http://nrr.georgetown.edu/NRR/nrrhome.htm
Androgen Receptor Mutations	http://www.androgendb.mcgill.ca
Estrogen Receptor Resource	http://nrr.georgetown.edu/Estrogen%20Receptor/ER-PAGE/Main.html
Glucocorticoid Resource	http://nrr.georgetown.edu/GRR/grr1.htm
PPAR Resource	http://www.cas.psu.edu/docs/CASDEPT/VET/jackvh/ppar/pparrfront.htm
Steroid Receptor Resource	http://nrr.georgetown.edu/NRR/srapr/srapr.html
Thyroid Receptor Resource	http://nrr.georgetown.edu/NRR/TRR/trrfront.html
Vitamin D Receptor Resource	http://vdr.bu.edu
Nuclear Receptor Database	http://receptors.ucsf.edu/NR/
Nuclear Receptor Structure Server	http://www.cmbi.kun.nl/NR/servers/html/
Nuclear Receptor Mutation Database	http://cmbipc60.cmbi.kun.nl:8080/cgi-bin2/nrmd/nrmd.py
Vitamin D Receptor Database	http://vdr.bu.edu/index.html
Photoreceptor Nuclear Receptor Database	http://www.retina-international.com/sci-news/nr2e3mut.htm

URL, uniform resource locator; PPAR, Peroxisome Proliferator-Activated Receptor.

Resource, and Thyroid Receptor Resource are at different stages of development, with collections of facts, sources for clones and antibodies, maps/sequences of receptor expression vectors, phenotypes of mutations, and protein alignments.

The Nuclear Receptor Database (NuclearRDB; http://receptors.ucsf.edu/NR/) (*24*) is formatted similarly to the GPCRDB described earlier and provides a resource for nucleotide and protein sequence information (from multiple species), multiple sequence alignments, phylogenetic trees, computationally predicted nuclear localization signals, 2-D and 3-D structures and alignments, and mutation information; information on binding partners (i.e., ligands, co-modulators) is in development. The Nuclear Receptor Structure Servers (http://www.cmbi.kun.nl/NR/servers/html/) are computational tools contained within NuclearRDB and are designed as a versatile and flexible way to provide structure information, such as torsion angles and relative surface exposure of the receptor molecule, and also serve as tools to calculate contacts of receptors with ligands and drugs (*25*).

The Nuclear Receptor Mutation Database (NRMD; http://cmbipc60.cmbi.kun.nl:8080/cgi-bin2/nrmd/nrmd.py) (*26*) provides a searchable database of mutation and mutagenesis information on nuclear receptors from multiple species. This database integrates data from SwisProt, NuclearRDB, the Vitamin D Receptor Database (VDR; http://vdr.bu.edu/index.html), the Photoreceptor Nuclear Receptor database (http://www.retina-international.com/sci-news/nr2e3mut.htm), and the GRR discussed earlier. The VDR contains ligand-binding features by chemical modification, site-directed mutagenesis, and homology-extension modeling. NRMD presently contains 893 mutations on 54 nuclear receptors.

3.3. Other Receptor Family Databases

The goal of the Integrated Receptor Database (IRDB; http://impact.nihs.go.jp/RDB.html) (*27*) is to provide "one-stop shopping" on receptor data and contains information on structural data and binding sites, cell signaling pathways triggered by ligand binding, and binding affinities. It also contains a viewer to represent information on endocrine disruptors and drug design, and information on single nucleotide polymorphisms (SNPs). Still in development, IRDB contains information on 1780 receptors, 250 DNA-binding sites, 170 ligand-binding sites, and 410 3-D structures. For example, a query of the human delta opioid receptor will provide 2-D structural information and the drug (morphine) with which the receptor binds.

The Human Plasma Membrane Receptome (http://receptome.stanford.edu/HPMR/home.asp) (*28*) combines text- and sequence-based tools for studying plasma membrane receptors in the human genome and provides gene information, summarizes ligand interactions, and links to literature, sequence, expression, and domain databases. It includes information on 7-transmembrane receptors, T-cell receptors, tumor necrosis factor (TNF) receptors, cytokine receptors, netrin receptors, integrins, plexins, and natriuretic peptide receptors.

The Ligand Gated Ion Channel Database (http://www.ebi.ac.uk/compneur-srv/LGICdb/LGICdb.php) (*29*) contains 513 entries from three superfamilies of extracellularly activated ligand-gated ion channel subunits. The cys-loop superfamily (nicotine receptor, $GABA_A$ and $GABA_C$ receptors, glycine receptors, $5\text{-}HT_3$ receptors, and some glutamate activated anionic channels) consists of five homologous subunits. The ATP-gated channels (ATP2x receptors) are composed of three homologous subunits. Finally, the glutamate-activated cationic channels (*N*-methyl-D-aspartate [NMDA] receptors, α-amino-3-hydroxy-5-methyl-4-isoxazole propionate [AMPA] receptors, Kainate receptors, and so on) are each assembled as four homologous subunits. Because of the lack of evolutionary relationship, information on these three superfamilies are included in separate databases. This family of databases provides sequence data, alignments, and phylogenic relationships using sequence alignments.

A number of mutations are known in the low-density lipoprotein (LDL) receptor (LDLR) to cause familial hypercholesterolemia (FH). The LDLR Database (http://www.umd.necker.fr) (***30***) compiles information on 840 LDLR mutations providing sequence information, domain location of mutation, proband ethnicity, occurrence in population, functional class, and clinical status of mutations (***31***). Additional LDLR mutation data can be accessed at http://www.ucl.ac.uk/fh.

IUPHAR (http://iuphar.org) details the molecular, biophysical, and pharmacological properties of mammalian sodium, calcium, and potassium channels, cyclic nucleotide-modulated ion channels, and transient receptor potential (TRP) channels in addition to GPCRs (*see* **Subheading 3.1.**). Information includes historical and current nomenclature systems, molecular structure variations within and across species, biophysical properties for functional subunits, pharmacological descriptions regarding agonists and antagonists, affinity values from radioligand-binding assays, and physiological and pathological roles.

4. Additional Databases That Contain Receptor–Ligand Information

With the multitude of published receptor–ligand and mutagenesis studies, databases cataloging various properties of receptor and ligands and experimental evidence of receptor–ligand interactions can provide a helpful resource. Several such resources exist and are summarized below.

The K_i Database (http://kidb.bioc.cwru.edu/) (***32***) is an "information warehouse" for published and internally derived affinity (K_i) values for GPCRs, ion channels, transporters, and enzymes. It contains three tools: (1) a graphing tool which allows for color-coded output of multiple ligands and multiple receptors and uses an algorithm (which collates all the relevant K_i values, averages them, culls the "outliers", and outputs a color-coded representation) to generate an average K_i value for a particular receptor–ligand pair; (2) a receptor-mining tool that allows the user to select two receptors and a source to compare K_i values; and (3) a ligand-selectivity tool that allows one to search for ligands with specified affinities for a particular molecular target so as to determine the selectivity of such compounds. Chemical structures are also linked for many compounds.

The Database of Ligand–Receptor Partners (DLRP; http://dip.doe-mbi.ucla.edu/dip/DLRP.cgi) (***33***) is a subset of the Database of Interacting Proteins (DIP; http://dip.doe-mbi.ucla.edu/) (***34***) that catalogs experimentally determined interactions between ligand–receptor complexes. DLRP combines information from a variety of sources to create a single, consistent set of receptor–ligand interactions. The data stored within the DIP and DLRP databases are curated, both manually by expert curators and also automatically using computational approaches. Information on various receptors and their ligands, such as the bone morphogenetic protein receptor and the chemokine receptors, are included in DLRP.

PreBIND (http://www.blueprint.org/products/prebind/prebind.html) is a data-mining tool to locate biomolecular interaction information in the scientific literature. Querying by the name or accession number of a protein returns a list of potentially interacting proteins. PreBIND can be used in conjunction with the Biomolecular Interaction Network Database (BIND; http://bind.ca), which is a collection of records documenting molecular interactions: molecules that associate with each other to form interactions, molecular complexes that are formed from one or more interaction(s), and pathways that are defined by a specific sequence of two or more interactions. The contents of BIND include high-throughput data submissions and hand-curated information gathered from the literature and created for interactions that have been shown experimentally and published in at least one peer-reviewed journal. BIND includes references to articles with experimental evidence that supports or disputes the associated interaction. All information is stored in BIND database records that are freely available through a web interface so as to allow users to query, view, and submit records. A query of a ligand (e.g., dopamine) links the user to its receptors (e.g., D5 receptor), the biological process this interaction regulates (e.g., GPCR signal transduction), and type of experimental evidence (e.g., "affinity"), as well as links to PubMed and National Council on Biotechnology Informtion (NCBI) sequences for interacting receptors.

Other smaller, but informative, websites for ligand–receptor–binding information also exist. The Protein Ligand Database (PLD; http://www-mitchell.ch.cam.ac.uk/pld/index.html) is a resource containing biomolecular data, including binding energies, Tanimoto ligand similarity scores, and protein sequence similarities of protein–ligand complexes; the PLD (v 1.3) currently has data on 485 protein–ligand complexes. The Binding database (BindingDB) (http://www.bindingdb.org/bind/stat.jsp) contains measured binding affinities for biomolecules, genetically or chemically modified biomolecules, and synthetic compounds. BindingDB currently contains data generated by isothermal titration calorimetry (ITC) and enzyme inhibition methods; other techniques will be included in the future. Information on agonists and antagonists of a subset of receptors can be found at http://www.neuro.wustl.edu/neuromuscular/lab/molecule.htm.

5. Ligand–Receptor Pathway Resources

Because ligand binding to a receptor is the first step in pathways that regulate biological response, assessing data from signal transduction pathways can indicate "downstream" responses to receptor–ligand interaction, and is another approach for obtaining receptor–ligand-binding information.

The Alliance for Cellular Signaling (AFCS; http://www.cellularsignaling.org/) has chosen a limited number of cellular systems to explore signal transduction pathways. Of particular relevance to the topic of ligand–receptor binding is the "Ligand Screen Data" contained in the AFCS. The Ligand Screen is a strategy for detecting the inputs and, in time, the combinations of inputs that are most relevant to regulation of the behavior of the cells under study. The initial goals are: (1) to determine which ligands give functionally unique responses; and (2) to define the combinations of ligands whose interaction is not simply energetically additive. Definition of the extent of the interactions among ligands is a key goal. A long-term goal is to analyze and quantitate the combinations of inputs that display the most robust interactions, because such interactions help define the level of complexity of the signaling network. In addition, "The Molecule Pages" is a database of keys facts about proteins involved in cellular signaling. It currently covers more than 3000 proteins, including receptors. For each of these, the database provides a large amount of "automated" data, collected from numerous other on-line resources and updated monthly. These data include names, synonyms, sequence information, biophysical properties, domain and motif information, protein family details, structure and gene data, the identities of orthologs and paralogs, and Basic Local Alignment Search Tool (BLAST) results. Information is provided for more than 800 proteins as "Mini Molecule Page" summaries composed by invited expert authors.

Another means to evaluate pathways that are activated by particular receptor systems is the Gene MicroArray Pathway Profiler (GenMAPP), which is a pathway-oriented approach for analyzing genome-scale experiments. GenMAPP is a freely distributed software package that can be downloaded at http://www.GenMAPP.org (***35***). This database allows the user to rapidly analyze and group large amounts of gene expression data by mapping changes in specific genes onto known biochemical or signaling pathways. In settings in which gene expression data are collected following treatment with particular ligands or activation of specific receptor systems, GenMAPP can provide functional information on patterns of response to ligand–receptor binding.

Notes: *The Database Issue* of *Nucleic Acid Research*, which appears as the first issue each year (since 1996), is a good resource for recent updates on publicly available databases.

Many databases and computational websites referenced in this chapter are rapidly evolving. The URLs, number of data entries, available tools, and version numbers reflect information available December 31, 2004.

Acknowledgment

The authors thank Phil Bourne for his helpful comments.

References

1. Rana, B. K. and Insel, P. A. (2001) Useful G-protein-coupled receptor websites. *Trends Pharm. Sci.* **22**, 485–486.
2. Rana, B. K. and Insel, P. A. (2002) G-protein-coupled receptor websites. *Trends Pharm. Sci.* **23**, 535–536.
3. Berman, H. M., Westbrook, J., Feng, Z., et al. (2000) The Protein Data Bank. *Nucleic Acids Res.* **28**, 235–242.
4. Brooijmans, N. and Kuntz, I. D. (2003) Molecular recognition and docking algorithms. *Ann. Rev. Biophys. Biomol. Struct.* **32**, 335–373.
5. Ewing, T. J. A, and Kuntz, I. D. (1997) Critical evaluation of search algorithms used in automated molecular docking. *J. Comput. Chem.* **18**, 1175–1189.
6. Hendlich M., Bergner A., Gunther J., and Klebe G. (2003) Relibase: design and development of a database for comprehensive analysis of protein-ligand interactions. *J. Mol. Biol.* **326**, 607–620.
7. Roche, O., Kiyama, R., and Books, C. L. (2001) Ligand-protein database: linking protein-ligand complex structures to binding data. *J. Med. Chem.* **44**, 3592–3598.
8. Ewing, T. J., Makino, S., Skillman, A. G., and Kuntz, I. D. (2001) DOCK 4.0: search strategies for automated molecular docking of flexible molecule databases. *J. Comput. Aid. Mol. Des.* **15**, 411–428.
9. Morris, G. M., Goodsell, D. S., Halliday, R. S., et al. (1998) Automated docking using a lamarckian genetic algorithm and empirical binding free energy function. *J. Comput. Chem.* **19**, 1639–1662.
10. Chen, X., Ji, Z. L., Zhi, D. G., and Chen, Y. Z. (2002) CLiBE: a database of computed ligand binding energy for ligand-receptor complexes and its potential use in the analysis of drug binding competitiveness. *J. Comput. Chem.* **26**, 661–666.
11. Chen, Y. Z., Gu, X. L., and Cao, Z. W. (2001) Can an optimization/scoring procedure in ligand-protein docking be employed to probe drug-resistant mutations in proteins? *J. Mol. Graph. Model.* **19**, 560–570.
12. Chen, Y. Z. and Ung, C. Y. (2001) Prediction of potential toxicity and side effect protein targets of a small molecule by a ligand-protein inverse docking approach. *J. Mol. Graph Model.* **20**,199–218.
13. Stuart, A. C., Ilyin, V. A., and Sali, A. (2002) LigBase: a database of families of aligned ligand binding sites in known protein sequences and structures. *Bioinformatics* **18**, 200–201.
14. Wallace, A. C., Laskowski, R. A., and Thornton, J. M. (1995). LIGPLOT: a program to generate schematic diagrams of protein-ligand interactions. *Prot. Eng.* **8**, 127–134.

15. Lichtarge, O., Bourne, H. R., and Cohen, F. E. (1996). An evolutionary trace method defines binding surfaces common to protein families. *J. Mol. Biol.* **257**, 342–358.
16. Horn, F., Bettler, E., Oliveira, L., Campagne, F., Cohen, F. E., and Vriend, G. (2003) GPCRDB information system for G protein-coupled receptors. *Nucl. Acids Res.* **31**, 294–297.
17. Beukers, M. W., Kristiansen, K., Ijzerman, A. P., and Edvardsen, O. (1999) TinyGRAP database: a bioinformatics tool to mine G protein-coupled receptor mutant data. *Trends Pharm. Sci.* **20**, 475–477.
18. Takeda, S., Kadowaki, S., Haga, T., Takaesu, H., and Mitaku, S. (2002) Identification of G protein-coupled receptor genes from the human genome sequence. *FEBS Lett.* **520**, 97–101.
19. Fredriksson, R., Lagerstrom, M. C., Lundin, L.G., and Schioth, H. B.(2003) The G-protein-coupled receptors in the human genome form five main families. Phylogenetic analysis, paralogon groups, and fingerprints. *Mol. Pharmacol.* **63**, 1256–1272.
20. Möller, S., Vilo, J., and Croning, M. D. R. (2001) Prediction of the coupling specificity of G protein coupled receptors to their G proteins. *Bioinformatics* **17**, S174–S181.
21. Qian, B., Soyer, O. S., Neubig, R. R., and Goldstein, R. A. (2003) Depicting a protein's two faces: GPCR classification by phylogenetic tree-based HMMs. *FEBS Lett.* **554**, 95–99.
22. Crasto, C., Marenco, L., Miller, P.L., and Shepherd, G.S. (2002) Olfactory Receptor Database: a metadata-driven automated population from sources of gene and protein sequences. *Nucl Acids Res.* 354–360
23. Martinez, E., Moore, D.D., Keller, E., Pearce, D., Vanden Heuvel, J.P., Robinson, V., Bottlieb, B., MacDonald, P., Simons, S. Jr., Sanchez, E., Danielsen, M. (1998) The Nuclear Receptor Resource: a growing family. *Nucl. Acids Res.* **26**, 239–241.
24. Horn, F., G. Vriend and F.E. Cohen (2001) Collecting and Harvesting Biological Data: The GPCRDB & NucleaRDB Databases. *Nucl Acids Res.* **29**, 346–349.
25. Bettler, E., Krause, R., Horn, F., Vriend, G. (2003) NRSAS: Nuclear Receptor Structure Analysis Servers. *Nucl. Acids Res.* **31**, 3400-3403.
26. Van Durme, J.J., Bettler, E., Folkertsma, S., Horn, F., Vriend, G. NRMD: Nuclear Receptor Mutation Database. *Nucl. Acids Res.* **31**, 331–333
27. Nakata, K., Takai-Igarashi, T., Nakano, T., Kaminuma, T. (2002) An integrated receptor database (IRDB). *Data Science Journ.* **1**, 140–145.
28. Ben-Shlomo, I., Yu Hsu, S., Rauch, R., Kowalski, H. W., and Hsueh, A. J. (2003) Signaling receptome: a genomic and evolutionary perspective of plasma membrane receptors involved in signal transduction. *Sci STKE.* **2003(187)**, RE9.
29. Le Novere, N. and Changeux, J.-P. (1999) The ligand gated ion channel database. *Nucl. Acids Res.* **27**, 340–342.
30. Varret, M., Rabes, J. P., Thiart, R., et al. (1998) LDLR Database (second edition): new additions to the database and the software, and results of the first molecular analysis. *Nucl. Acid Res.* **26**, 248–252.

31. Villeger, L., Abifadel, M., Allard, D., et al. (2002) The UMD-LDLR database: additions to the software and 490 new entries to the database. *Hum. Mutat.* **20**, 81–87.
32. Roth, B. L., Kroeze, W. K., Patel, S., and Lopez, E. (2000) The multiplicity of serotonin receptors: uselessly diverse molecules or an embarrassment of riches? *The Neuroscientist* **6**, 252–262.
33. Graeber, T. G. and Eisenberg, D. (2001) Bioinformatic identification of potential autocrine signaling loops in cancer using gene expression profiles. *Nat. Genet.* **29**, 295–300.
34. Xenarios, I., Salwinski, L., Duan, X. J., Higney, P., Kim, S., and Eisenberg, D. (2002) DIP: The Database of Interacting Proteins. A research tool for studying cellular networks of protein interactions. *Nucleic Acid Res.* **30**, 303–305.
35. Dahlquist, K. D., Salomonis, N., Vranizan, K., Lawlor, S. C., and Conklin, B. R. (2002) GenMAPP, a new tool for viewing and analyzing microarray data on biological pathways. *Nat. Genet.* **31**, 19–20.

2

Identification of Orphan G Protein-Coupled Receptor Ligands Using FLIPR® Assays

Nicola M. Robas and Mark D. Fidock

1. Orphan GPCRs

G protein-coupled receptors (GPCRs) make up the largest and most diverse family of transmembrane proteins and respond to a wide variety of stimuli including biogenic amines, peptides, bioactive lipids, hormones, and light (***1,2***). Agonist binding to these receptors activates intracellular signalling events mediated by G proteins, such as modulation of intracellular cyclic adenosine monophosphate (cAMP) levels or Ca^{2+} mobilization. To date, there are approx 250 characterized nonsensory GPCRs and a further 140 genes predicted to be GPCRs for which the endogenous or natural ligand is unknown—the "orphan" GPCRs (oGPCRs) (***3–5***). Historically, GPCRs, especially those in the aminergic receptor subfamily, have proved amenable to the design of synthetic agonists and antagonists of their activity. Of the top-selling prescription drugs in 2002, more than 33% act through GPCRs and provide greater than $25 billion in worldwide pharmaceutical sales. Therefore, considerable effort has been made to identify cognate ligands for oGPCRs and functionally characterize these receptors in order to elucidate their physiological and therapeutic relevance.

1.1. Promiscuous and Chimeric G Proteins

GPCRs exert their effects via activation of a variety of signaling pathways, mediated by the interaction of the receptor with its cognate G protein. There are four main families of G proteins whose functions are determined by their α

From: *Methods in Molecular Biology, vol. 306: Receptor Binding Techniques: Second Edition*
Edited by: A. P. Davenport © Humana Press Inc., Totowa, NJ

subunit: $G\alpha_s$ activates adenylate cyclase, $G\alpha_i$ inhibits adenylate cyclase, $G\alpha_q$ activates phospholipase C, and $G\alpha_{12}$ has diverse signalling characteristics including modulation of Na^+/H^+ exchange and c-Jun N-terminal kinase (JNK) activation (***6***). For an oGPCR, not only is the ligand unknown, but also its G protein partner and the associated signaling cascade. One of the most successful high-throughput methods for oGPCR ligand screening is the measurement of changes in intracellular Ca^{2+} as a result of receptor-mediated phospholipase Cβ1 (PLC) activation (***7–11***) (**Table 1**). However, only a subset of naturally occurring G proteins signal through the PLC cascade; therefore, a mechanism is needed to channel a spectrum of downstream signaling pathways to a single measurable end point. To this end, the "promiscuous" G proteins e.g., $G\alpha_{15}$ or $G\alpha_{16}$, together with G protein chimeras, such as $G\alpha_{qi5}$ and $G\alpha_{qs5}$, are widely used. $G\alpha_{15}$ and $G\alpha_{16}$ are naturally occurring G proteins with the ability to couple to receptors which would normally signal via an alternative pathway (***12***). Using this characteristic, it is possible to "force" a receptor to respond to an agonist via PLC activation, thus considerably broadening the range of receptors that will give a measurable calcium mobilization response. Chimeric G proteins consist of $G\alpha_q$ in which the C-terminal five amino acids of this subunit are replaced by corresponding amino acids from the adenylate-cyclase linked G_i or G_s subunit to generate $G\alpha_{qi5}$ and $G\alpha_{qs5}$ respectively (***13***). Thus, these chimeras allow most G_i or G_s coupled receptors to signal via elevation of intracellular Ca^{2+}. However, it should be noted that although the majority of GPCR linked pathways can be manipulated in this way, this system is not universally applicable (***14***) and platforms utilizing cAMP response elements can be considered for putative G_i and G_s receptors that are unresponsive in a Ca^{2+} assay.

1.2. Fluorescent Calcium-Sensitive Dyes

Elevation of cytoplasmic Ca^{2+} resulting from receptor-coupled release from intracellular stores can be detected by using calcium-sensitive dyes such as Fluo-3 acetoxymethyl (AM) and Fluo-4 AM (Molecular Probes, www.Probes.com) which exhibit an increase in fluorescent intensity upon binding to Ca^{2+} (***15***). Incubation of the cells with the cell permeable indicator allows "loading" of the cytoplasm, and cleavage of the AM ester moiety by cytoplasmic esterases prevents the active dye from diffusing out of the cell. Because the AM form has a low aqueous solubility, a dispersion agent, e.g., Pluronic F-127, is used to facilitate cell loading. For some cell types, e.g., Chinese hamster ovary (CHO) cells, the inclusion of an anion exchange inhibitor, such as probenecid, is required for efficient cell loading (***16***). The Fluo-4 dye formulation requires the cells to be washed prior to processing to remove residual extracellular dye that can increase background signals. Recently, "no wash" dye formulations have

Table 1
Examples and Year of Publication of Orphan Receptor–Ligand Pairings That Used Ca^{2+} Flux As the Assay Readout

Ligand	Receptor	Year	Reference
Cortistatin	MrgX2	2003	*7*
Sphingosine 1-phosphate	GPR3/6/12	2002	*8*
BAM22 and related fragments	SNSR3/4	2002	*9*
KISS-1	GPR54	2001	*10*
Melanin concentrating hormone	MCH2	2001	*11*

*These citations are illustrative and not comprehensive as in some cases several groups identified the same receptor–ligand pair.

been developed and are commercially available, e.g., fluorometric imaging plate reader (FLIPR®) Calcium 3 assay reagent (Molecular Devices). The main advantages these provide over the protocols in which wash steps are required is an increase in throughput capability and a reduction in the stress put on the cells, especially for fragile cells or those that are weakly adherent. We have found both Fluo-4 and Calcium 3 reagents to have excellent signal sensitivity. However, when using Calcium 3 reagent we have found high background signals with lipid ligands and greater variability between cell types than when using Fluo-4.

1.3. Fluorometric Imaging Plate Reader

The protocols detailed here describe the use of the FLIPR 96-well microplate system for the measurement of intracellular calcium levels. The FLIPR hardware contains optic, liquid-handling, and temperature-control systems together with data collection and analysis software. The FLIPR comprises a 96-well pipettor that simultaneously adds compounds to a microplate containing the cell type to be tested. The cell monolayer is then excited with an argon laser, and the resulting fluorescence change in response to compound treatment is detected by a charge-coupled device (CCD) camera in real time (*see* www.moleculardevices.com for more information). When monitoring an agonist treatment using Fluo-4 or Calcium 3 assay reagent, a typical assay can be run in approx 4 min per plate.

1.4. Selection of Ligand Library

The success rate of ligand identification for an orphan receptor will depend on a number of factors including ligand library selection and the concentrations used. We commonly use a 10 μ*M* concentration for small molecule ligands and bioactive lipids, and a 1 μ*M* concentration for peptides. The choice of compound library will depend on whether a natural or a synthetic ligand is required. A selection of commercially available GPCR ligand libraries (96-well format) is listed in the "Materials" section (**Subheading 2.3.**). An alternative source of ligands is the use of high-performance liquid chromatography (HPLC) fractions prepared from tissue extracts (***17***).

2. Materials

2.1. Transient Transfection of HEK293 (Human Embryonic Kidney) Cells

1. Microbiological Safety Cabinet (class II).
2. Microscope.
3. CO_2 incubator set at 37°C with humidified 5% CO_2/95% air e.g., Hereaus.
4. Hemocytometer (Sigma, www.sigmaaldrich.com).
5. Rechargeable pipetman.
6. 225-cm^3 Flasks, vented cap (Costar, www1.fishersci.com).
7. HEK293 cells (Human embryonic kidney) (Invitrogen, www.invitrogen.com). Note: alternative cell lines such as CHO or COS-7 can also be used.
8. Dulbecco's modified Eagle medium (DMEM) + 10% fetal calf serum, 2 m*M* L-glutamine, 25 m*M* HEPES, 1X MEM nonessential amino acids.
9. Serum free minimal media, e.g., OptiMEM (Invitogen).
10. Cationic lipid transfection agent, e.g., Lipofectamine Plus (Invitrogen).
11. Mammalian expression constructs for $G\alpha_{15}$ (Genbank: AF493904, plasmid available from Molecular Devices) and $G\alpha_{qi5}$ (plasmid available from Molecular Devices).

2.2. FLIPR Assay

1. 1X Trypsin 0.25 mg/mL (Invitrogen).
2. Phosphate-buffered saline (PBS) without calcium or magnesium.
3. 96-well black, clear-bottomed microplate, sterile (Costar).
4. Probenecid (Sigma).
5. Type B 96-well black FLIPR tips (Molecular Devices).
6. FLIPR (Fluorometric Imaging Plate Reader) (Molecular Devices).
7. Sterile reservoirs (Costar).
8. 1 *M* sodium hydroxide.

2.2.1. Fluo-4 AM Protocol

1. Fluo-4 AM (Molecular Probes).
2. Pluronic F-127 (Molecular Probes).
3. Dimethyl sulfoxide (DMSO).
4. 1X Hank's balanced salt solution (HBSS).
5. Bovine serum albumin (BSA).

2.2.2. Calcium 3 Assay Reagent (No Wash)

1. FLIPR Calcium 3 assay reagent (Molecular Devices).
2. 1X HBSS.
3. 1 *M* HEPES.

2.3. GPCR Ligands

1. RBI Library of Pharmacologically Active Compounds (LOPAC): 640 small molecule ligands (agonists + antagonists) for known GPCRs (Sigma).
2. Prestwick chemical library: 880 pharmacologically active compounds (Prestwick Chemical, Inc.; www.prestwickchemical.com).
3. Prestwick peptide library: 240 known peptide ligands (Prestwick Chemical, Inc.).
4. Biomol lipid library: 203 bioactive lipids (Biomol; www.Biomol.com).
5. Biomol Orphan Ligand Library: 84 compounds with defined or putative biological activity whose protein-binding partners are unknown (Biomol).

3. Methods

The methods in the Hsubsequent sections use HEK293 cells and the Lipofectamine Plus transfection reagent. Alternative standard cell lines, e.g., CHO, COS-7, and alternative cationic lipids, can also be used.

Note: all cell handling to be carried out in a Microbiological Safety Cabinet Class II.

3.1. Transient Transfection of HEK293 Cells

1. Cells grown to 60–80% confluency in 225cm^2 vented flask.
2. Prewarm OptiMEM to 37°C.
3. Prepare solution A: 15 μg oGPCR plasmid DNA, 5 μg $G\alpha_{15}$ plasmid, 5 μg Gq_{i5} plasmid, 90 μL Plus reagent, 2.25 mL OptiMEM.
4. Incubate Solution A at room temperature for at least 15 min.
5. Meanwhile, prepare Solution B: 45 μL Lipofectamine (cationic lipid), 2.25 mL OptiMEM.
6. Combine solutions A and B. Mix gently. Incubate at room temperature for 15 min.
7. Add 15 mL OptiMEM.
8. Remove cells from incubator. Wash in 15 mL OptiMEM and then aspirate off the liquid.
9. Add the DNA/lipofectamine/OptiMEM mix.
10. Incubate at 37°C for 4–5 h.
11. Aspirate the transfection mix and add 40 mL full growth media.
12. Return to CO_2 incubator set at 37°C with humidified 5% CO_2/95% air.
13. Prepare mock-transfected cells alongside receptor transfected cells in order to determine responses resulting from the cellular background (*see* **Note 1**).

3.2. FLIPR Assay

3.2.1. Cell Preparation

1. Twenty-four hours posttransfection, wash cells in 10 mL prewarmed PBS and then aspirate off the liquid.
2. Add 2 mL trypsin. Incubate for 1–2 min at 37°C.
3. Gently tap flask to detach the cells. Add 4 mL growth media + serum. Count cells using a hemocytometer.
4. Seed cells into black, clear-bottomed 96-well plates at a density of 5×10^4 cells per well in a 100 μL volume (*see* **Notes 2–4**).
5. Culture cells for a further 24 h.

3.2.2. Compound Preparation

1. Compounds diluted to working concentration in round-bottomed, 96-well plates (*see* **Note 5**).
2. Compound working stocks are made up at 4X concentration for Fluo-4 assay and 5X concentration for the FLIPR Calcium 3 assay, to account for the dilution that occurs on addition to the cell plate; e.g., in the Fluo-4 assay, for a screening concentration of 10 μ*M*, compound addition plate contains 40 μ*M* stock.
3. Small molecule ligands and bioactive lipids are screened at a final assay concentration of 10 μ*M* and peptides at 1μ*M*.
4. Peptides are diluted to the appropriate concentration in HBSS containing 0.1% BSA. All other compounds are diluted in HBSS where possible (*see* **Note 6**).

3.2.3. Preparation of Fluo-4 Loading Dye and Loading the Cells (48 h Posttransfection)

Note: keep loading dye protected from direct light (*see* **Note 7**).

1. Prepare 100X probenecid stock (250 m*M*) by dissolving 0.71 g probenecid in 5 mL of 1 *M* sodium hydroxide. Make up to 10 mL with PBS. Prepare fresh on the day of the assay.
2. To make up loading dye solution for one assay plate, thaw one vial (50 μg) Fluo-4 and resuspend in 20 μL DMSO.
3. Add 20 μL 20% pluronic F-127 and mix.
4. Add dye/pluronic mixture to 11 mL serum free growth medium.
5. Add 1.1 mL 100X probenecid stock.
6. Remove growth medium from the cells and replace with 100 μL warm (37°C) loading dye solution (*see* **Note 8**).
7. Incubate for 1 h at 37°C in 5% CO_2/95% air.
8. Wash cells three times with 150 μL of wash buffer per well (HBSS/2.5 m*M* probenecid, pH 7.4) (*see* **Note 9**). When using peptide ligands include 0.1% BSA.
9. Incubate at room temperature for 15 min prior to processing within the FLIPR.

Table 2
FLIPR® Program Settings

Laser setting	0.6 W
Exposure length	0.4 s
Addition volume	50 µL
Pipettor height	180 µL (Fluo-4) or 230 µL (Calcium 3)
Addition speed	35 µL/s
Addition start	After sample 10 (after 10 s)
Read time (total = 2 min)	60 samples every 1 s followed by 20 samples taken every 3 s

3.2.4. Preparation of Calcium 3 Assay Reagent ("No Wash") and Loading of Cells

1. Prepare 100X probenecid stock (250 mM) by dissolving 0.71 g probenecid in 5 mL 1 *M* sodium hydroxide. Make up to 10 mL with PBS. Prepare fresh on the day of the assay.
2. To prepare the reagent buffer, pipet 10 mL of 10X Hanks Balanced Salt Solution, 2 mL of 1 *M* HEPES buffer solution, and 1 mL of 100X probenecid (final in-well concentration of 2.5 m*M*), into 86 mL cell culture treated water.
3. Thaw one vial FLIPR Calcium 3 assay reagent and equilibrate to room temperature.
4. Dissolve contents of one vial completely in 10 mL of reagent buffer and then add to the remaining buffer. Adjust pH to 7.4 and adjust volume to 100 mL with water.
5. Remove cell plates from incubator.
6. Add an equal volume (100 µL) of assay reagent to each well. Growth medium does not have to be removed. (*see* **Note 8**).
7. Incubate cell plates for 1 h at 37°C/5% CO_2 and then equilibrate to room temperature for 10 min prior to reading on the FLIPR.

3.2.5. Reading the Assay Plates on the FLIPR

Program the FLIPR to take readings as detailed in **Table 2** (**Notes 10–11**).

3.2.6. Data Analysis

Results are displayed as graphs in 96-well format. An agonist-dependent, receptor-mediated response is characterized by a distinctive calcium signal which returns to the baseline level (**Fig. 1A**). Care should be taken to identify non-receptor-mediated signals (**Fig. 1B**) such as those produced by calcium ionophores or compounds that disrupt the lipid bilayer. A numerical value representing the response can be obtained by exporting the statistics of each curve. The statistics most commonly used are Max-Min and Sum (area under the curve).

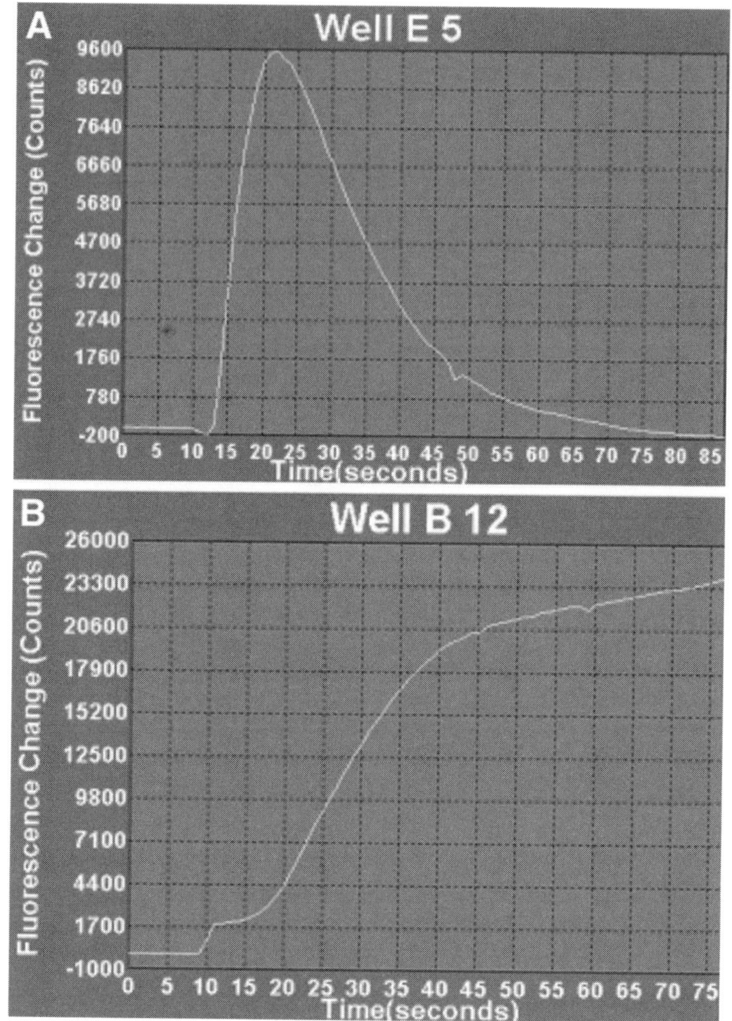

Fig. 1. (**A**) Example of a receptor-mediated fluorescence/calcium response characterized by a sharp peak within seconds of agonist addition (compound added at 10 s), followed by a return to baseline. (**B**) Example of a non-receptor-mediated fluorescence/calcium response characterized by slow onset and not returning to baseline.

Subtraction of the mock Ca^{2+} signal from the oGPCR transfected signals will identify any responses caused by receptors endogenous to the cell line used.

4. Notes

1. The transfection efficiency of the cell line and protocol can be checked by transfecting with a β-galactosidase reporter to estimate the percentage of cells that

have taken up the expression plasmid (β-galactosidase assay system, Invitrogen). Alternatively, a parallel transfection of a tagged receptor can be used.

2. Fragile or weakly adherent cells (e.g., HEK293) are best seeded on poly D-lysine coated plates, especially when using a protocol that includes wash steps.
3. To reduce well-to-well and plate-to-plate variability, a Multidrop dispenser (ThermoQuest) can be used for addition of cells to the assay plate.
4. On the day of the assay, the cells should be 90–100% confluent in the microplate. The 96-well FLIPR reads fluorescence across the middle of the well. Therefore, any spaces in the monolayer in this region will result in signal reduction.
5. When using round- or V-bottomed plates for compound preparation, at least 10 μL dead volume is required for the FLIPR liquid handling system (50 μL for flat-bottomed plates).
6. When testing compounds which require DMSO for solubilization, check the DMSO tolerance of your assay—most cell based screens are sensitive to DMSO >1%.
7. Do not expose the loading dye to direct light. When loading the cells, turn off the light in the tissue culture cabinet. When equilibrating the plates to room temperature, keep them covered.
8. When testing multiple plates, stagger the dye loading step at 5-min intervals (if carrying out a single compound addition) so that each plate is incubated for approx 1 h in loading dye. Shorter incubation times can affect sensitivity because of insufficient dye loading. Longer incubation times can increase background levels and affect cell viability. Signal stability can be tested by using different load times and incubation temperatures (room temperature vs 37°C).
9. Prior to carrying out an experiment, test the wash protocol on your cell type of choice and then check the monolayer under the microscope to ensure cells are not being dislodged.
10. The pipettor height should be above the volume of the loaded cells, but below the total volume once the compound is added. This prevents loss of compound through splashes on the sides of the well.
11. Prior to placing the cell plates in the FLIPR, wipe the bottom with an antistatic cloth to remove dust or fingerprints that may affect the signal.

References

1. Marinissen, M.J., and Gutkind, J.S. (2001) G protein-coupled receptors and signalling networks: emerging paradigms. *Trends Pharmacol. Sci.* **22(7),** 368–376.
2. Baldwin, J.M. (1994) Structure and function of receptors coupled to G proteins. *Curr. Opin. Cell. Biol.* **6(2),** 180–190.
3. Marchese, A., George, S.R., Kolakowski, L.F., Lynch, K.R., and O'Dowd, B.F. (1999) Novel GPCRs and their endogenous ligands: expanding the boundaries of physiology and pharmacology. *Trends Pharmacol. Sci.* **20(9),** 370–375.
4. Wilson, S., Bergsma, D.J., Chambers, J.K.,et al. (1998) Orphan G protein-coupled receptors: the next generation of drug targets? *Br. J. Pharmacol.* **125(7),** 1387–1392.
5. Howard, A.D., McAllister, G., Feighner, S.D., et al. (2001) Orphan G protein-coupled receptors and natural ligand discovery. *Trends Pharmacol. Sci.* **22(3),** 132–140.

6. Pangalos, M.N. and Davies, C.H. (2002) *Understanding G Protein-Coupled Receptors and Their Role in CNS*. Oxford University Press, Oxford, UK: 63–86.
7. Robas, N., Mead, E. and Fidock, M. (2003) MrgX2 is a high potency cortistatin receptor expressed in dorsal root ganglion. *J. Biol. Chem.* **278(45)**, 44,400-44,404.
8. Uhlenbrock, K., Gassenhuber, H., and Kostenis, E. (2002) Sphingosine 1-phosphate is a ligand of the human gpr3, gpr6 and gpr12 family of constitutively active G protein-coupled receptors. *Cell Signal* **14(11)**, 941–953.
9. Lembo, P.M., Grazzini, E., Groblewski, T., et al. (2002): Proenkephalin A gene products activate a new family of sensory neuron-specific GPCRs. *Nat. Neurosci.* **5(3)**:201–209.
10. Ohtaki, T., Shintani, Y., Honda, S., et al. (2001) Metastasis suppressor gene KiSS-1 encodes peptide ligand of a G protein-coupled receptor. *Nature* **411**(6837), 613–617.
11. Hill, J., Duckworth, M., Murdock, P., et al. (2001) Molecular cloning and functional characterization of MCH2, a novel human MCH receptor. *J Biol Chem*, **276(23),** 20,125–20,129.
12. Offermanns, S. and Simon, M. (1995) $G\alpha_{15}$ and $G\alpha_{16}$ Couple a Wide Variety of Receptors to Phospholipase C. *J. Biol. Chem.,***270(25),** 15,175–15,180.
13. Milligan, G. and Rees, S. (1999) Chimeric Gα proteins: their potential use in drug discovery. *Trends Pharmacol. Sci.*, **20**, 118–124.
14. Kostenis, E. (2001) Is Gα16 the optimal tool for fishing ligands of orphan G protein-coupled receptors? *Trends Pharmacol. Sci.*, **22**, 560–564.
15. Minta, A., Kao, J.P., and Tsien, R.Y. (1989). Fluorescent indicators for cytosolic calcium based on rhodamine and fluorescein chromophores. *J. Biol. Chem.* **264(14),** 8171–8178.
16. Di Virgilio, F., Steinberg, T.H., Silverstein, S.C. (1989) Organic-anion transport inhibitors to facilitate measurement of cytosolic free Ca^{2+} with fura-2. *Methods Cell Biol.* **31**, 453–462.
17. Shimomura, Y., Harada, M., Goto, M., et al. (2002) Identification of neuropeptide W as the endogenous ligand for orphan G protein-coupled receptors GPR7 and GPR8. *J. Biol. Chem.* **277,** 35,826–35,832.

3

Quantitative Analysis of Orphan G Protein-Coupled Receptor mRNAs by TaqMan® Real-Time PCR

G2A and GPR4 Lysophospholipid Receptor Expression in Leukocytes and in a Rat Myocardial Infarction-Heart Failure Model

Stephen A. Douglas, Zhaohui Ao, Douglas G. Johns, Kristeen Maniscalco, Robert N. Willette, Lea Sarov-Blat, John P. Cogswell, Sheila Seepersaud, Paul Murdock, Klaudia M. Steplewski, and Lisa Patel

1. Introduction

Historically, the G protein-coupled receptor (GPCR) protein family has proven to be an extremely tractable target class *(1)*. It is estimated that approximately one-half of all drugs currently marketed exert their actions, either directly or indirectly, via GPCRs *(2)*. Given the potential commercial opportunities emanating from the identification of small molecule modulators of "novel" GPCRs (currently, GPCRs generate in excess of $25 billion per year in worldwide sales revenue *[3]*), it is not surprising that it is with great enthusiasm that both the pharmaceutical industry and academia move toward identifying novel members of this protein class.

Advances in recombinant DNA and screening technologies, along with the recent completion of the sequencing of the human genome, have marked the beginning of a new era in drug discovery, one in which putative, novel GPCRs are routinely identified following *in silico* searches of large DNA databases *(1)*. To this end, numerous cDNAs have been cloned that are believed to encode

From: *Methods in Molecular Biology, vol. 306: Receptor Binding Techniques: Second Edition*
Edited by: A. P. Davenport © Humana Press Inc., Totowa, NJ

for putative members of the GPCR protein family. What remains unclear, however, is which (if any) of these genes are of (patho)physiological significance.

1.1. Orphan G Protein-Coupled Receptors

It is estimated that of the approx 30,000 genes encoded by the human genome, approx 1000 are members of the GPCR family. Of this number, some 30% represent what are believed to constitute "viable" drug targets (i.e., "nonsensory" [olfactory, gustatory] receptors). Around half of this group of "druggable" GPCRs are what are known as "orphan" receptors, that is, receptors for which, to date, no known ligand has been defined *(4)*. Progress in "fostering" these orphan GPCRs has been impressive over the last decade, resulting in the identification of dozens of novel ligand–receptor targets *(1)*. However, the ability to select the most "attractive" orphan receptors from this ever-expanding list—namely those that are associated unambiguously with a given disease process—has been somewhat less tractable.

1.2. G2A and GPR4 Receptors

Two putative "orphan" targets, GPR4 *(5)* and G2A *(6)*, have recently been identified as G_i-coupled receptors for the naturally occurring bioactive lysophospholipids sphingosylphosphorylcholine (SPC) and/or lysophosphatidylcholine (LPC). These serum lipids are purported to be involved in the initiation and maintenance of a variety of pro-inflammatory processes within the mammalian vasculature. It is presumed that the pathological actions of these lysophospholipids (T/B-lymphocyte maturation, monocyte recruitment, macrophage activation and so on) are mediated by GPR4 and/or G2A *(7,8)*. To date, however, little is known about the relative expression of either G2A or GPR4 in cell- and tissue-based models of myocardial infarction and heart failure, diseases with associated profound, chronic inflammatory processes *(9,10)*. Understanding the relative regulation of G2A and GPR4 mRNA expression might go some way toward an understanding of which, if either, of these two lysophospholipid receptors is most likely to be involved in the etiology of cardiovascular diseases, such as, for example, myocardial infarction/heart failure.

1.3. Target Validation: Prioritizing Receptors of Interest

The ability to rank candidate drug targets *a priori*, that is, with no prior discriminating knowledge of their (patho)physiological significance, remains a major obstacle in the "target validation" process today *(1,11)*. Part of this validation process centers on the understanding of which genes are likely to be associated with specific disease processes of interest and which are not. Although target validation is clearly a complicated process, and one involving extensive, detailed investigations using complex model systems (knock-out/

UNIVERSITY OF BRISTOL MEDICAL LIBRARY

transgenic rodents, forward and reverse genetics, siRNA, utilization of tool receptor inhibitors in standard disease models, gene association studies, and so on), an early step in this process centers on the evaluation of the tissue distribution of a given gene/protein ***(12)***. Preliminary examination of gene expression in "appropriate" cell types and tissues is of significant utility as an early step in a strategy for selecting "validated" (or perhaps, the *most* validated) targets. This is particularly so in the case of novel orphan receptors whose recent discovery precedes by several months the availability of suitable antibodies (for Western blot analysis, immunohistochemistry, and so on) or radio-/fluorescently labeled ligands (for quantitative receptor autoradiography, conventional binding studies, and so on).

The present study details the use of one such selection tool, TaqMan®-based real-time polymerase chain reaction (PCR). This method allows for the quantitative comparison of relative gene (G2A and GPR4) expression in both cell- and tissue-based model systems of cardiovascular disease.

1.4. TaqMan-Based Real-Time PCR As a Method for Quantitating GPCR mRNA Expression

Advances in mRNA quantitation technology now allow for the rapid quantification of tissue/disease-specific gene expression. One technique, TaqMan-based real-time PCR, is a rapid methodology for evaluation target mRNA distribution. Although this technique is frequently used to accurately quantitate the expression of novel "orphan" GPCRs (as a result of the lack of available pAbs, radioligands, and so on), it can just as readily be applied to determine the expression of any mRNA transcript (whether it encode for an orphan/liganded GPCR or an unrelated gene). Although the primary focus of those protocols described herein is to facilitate the quantitation of GPCR expression, the basic methodology can be used for any transcript (and can, indeed, also be successfully adapted for many other applications, including genotyping, diagnostics, pathogen quantitation, SNP/mutation screening, and so on).

Using a charge-coupled device (CCD) detector and a thermal cycler, "real-time" TaqMan protocols quantitate mRNA expression in 96- or 384-well formats by integrating PCR-based methodologies with laser scanning technology ***(13,14)***. During the PCR reaction, typically performed with an ABI Prism Model 7700 or 7900HT Sequence Detection System (SDS), a laser is used to excite one or more 5'-fluorescent dyes (FAM™, TET™, VIC™) present in/derived from a transcript-specific "TaqMan probe," a 20–30mer oligodeoxynucleotide designed to hybridize selectively to a homologous, internal sequence of the target gene cDNA (**Fig. 1A**). The TaqMan probe is designed to "nest" within a site-specific region of a target cDNA at some point between that recognized by the conventional forward and reverse PCR primers. When PCR amplification occurs (**Fig.

A Primers and probe annealing

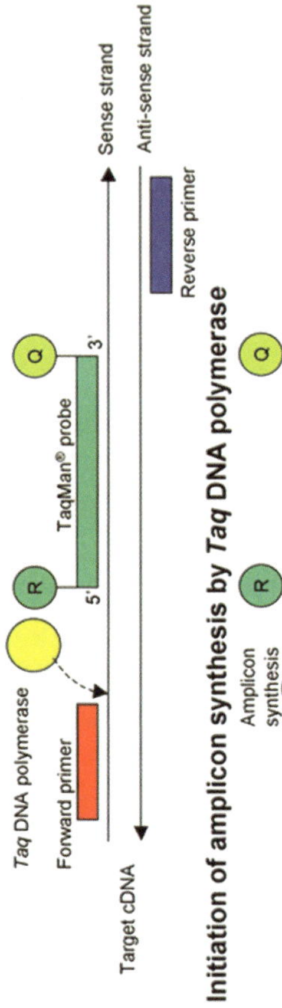

B Initiation of amplicon synthesis by *Taq* DNA polymerase

C Catabolism of TaqMan® probe during first strand amplicon extension (due to 5'-nuclease activity of *Taq* DNA polymerase) releases 5'-reporter dye from 3'-quencher/probe

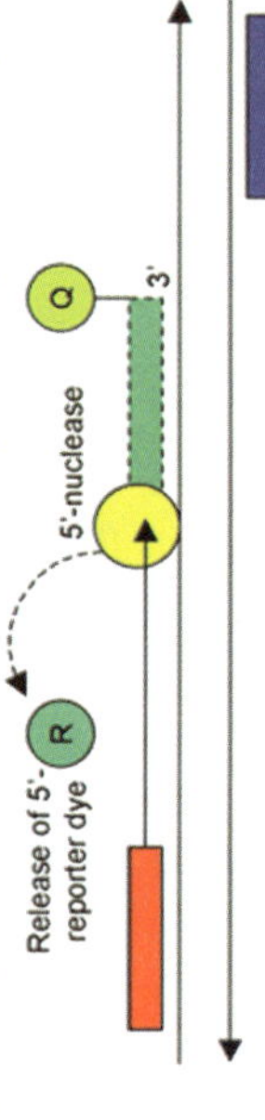

D Once "released" and no longer in close proximity to the 3'-quencher, the 5'-reporter dye emits light upon excitation by laser light, an event detected using a CCD camera

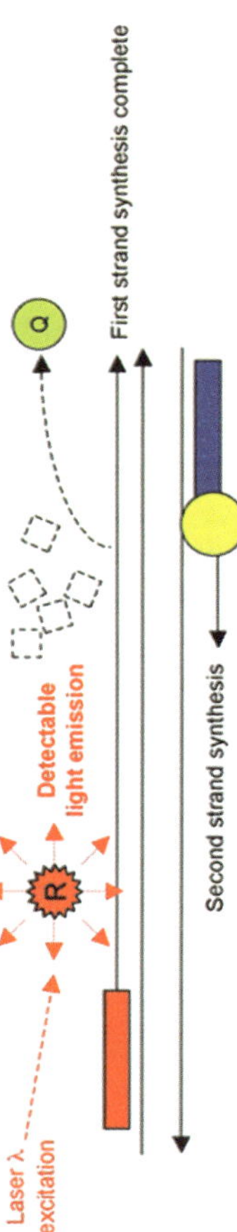

1B), the nesting TaqMan probe, labeled with a 5'-fluorescent reporter dye and a 3'-fluorescent quencher dye (TAMRA™), is degraded as a result of the 5'- to 3'-exonuclease activity of *Taq* DNA polymerase (hence this process is also referred to as a "fluorogenic 5'-nuclease assay"; **Fig. 1C**). As a result, the 5'-reporter dye is separated from the 3'-quencher portion of the TaqMan probe during PCR extension. This loss of proximity between the 5'-reporter and 3'-quencher results in an increase in fluorescence intensity of the reporter dye upon excitation (laser light; **Fig. 1D**). As such, the cycle-by-cycle amplification of PCR amplicon can be measured spectrophotometrically (during amplification, light emission increases proportionately in an exponential manner). As such, this simple and reliable methodology permits the rapid quantification of low levels of cDNA transcript in a "closed cap," automated system without the need for gel electrophoresis/densitometry. If so desired, the system can be multiplexed using multiple probe dyes facilitating the simultaneous evaluation of multiple target cDNAs. Assay sensitivity is estimated to be as low as approximately five copies of transcript/reaction with a wide dynamic range (in the order of 8 log units).

2. Materials

1. Standard reagents: typically, those found in any standard molecular biological laboratories (DNase-free or double distilled water, pipet tips, Eppendorf tubes, ethanol, chloroform, and so on; *see* **Note 1**)

Fig. 1. *(continued from facing page)* Schematic representation of the TaqMan®real-time polymerase chain reaction (PCR) process. (**A**) Under suitable conditions, target-specific primers and probes anneal to the denatured cDNA (sense and antisense) strands. (**B**) This facilitates complimentary DNA strand synthesis upon association with *Taq* DNA polymerase. Under these initial conditions, the intact TaqMan probe remains bound to the target DNA upstream of the polymerase. As such, when the probe is exposed to laser light, energy is transferred via flourescence resonance energy transfer (FRET) from the short-wavelength fluorophore on one 5'-end of the probe to the long-wavelength fluorophore on the 3'-end. The proximity of these two fluoroprobes results in the quenching the short-wavelength fluorescence. (**C**) However, as *Taq* DNA polymerase works its way along the target cDNA strand, it meets the TaqMan probe, which it is able to enzymatically degrade as a result of it's 5'-exonuclease activity. (**D**) Upon degradation, the FRET interaction is interrupted, leading to an increase in fluorescence from the short-wavelength 5'-fluorophore (and a decrease in fluorescence from the long-wavelength 3'-fluorophore). The resultant increase in fluorescence from the short-wavelength fluorophore, quantified by the optics of the ABI Prism® 7700 Sequence Detection System machine, is proportionate to the amount of PCR product generated. As such, fluorescence can, in turn, be used to back calculate the amount of template present in the original cDNA sample using standard curves. CCD, charge-coupled device.

2. Source of RNA: target cells or tissue.
3. ABI Prism Model 7700 (or 7900HT): sequence detection system (CCD detector/thermal cycler) required for TaqMan real-time PCR.
4. RNAlater (Qiagen, Ambion).
5. Tri reagent (e.g., Trizol, Gibco BRL, Tri reagent, Sigma).
6. RNase-Free DNase Set (Qiagen): dissolve DNase I (1500 Kunitz units) in 550 µL RNase-free water. Mix gently by inversion. Do not vortex. For long-term storage, remove stock solution from the glass vial, divide into single-use aliquots and store at –20°C for 9 mo. Thawed aliquots can be stored at 2–8°C for 6 wk. Do not refreeze after thawing.
7. RNeasy mini kit (Qiagen).
8. Ribogreen RNA quantitation reagent: 1 mL solution in dimethyl sulfoxide (DMSO).
9. 1X TE buffer: 10 m*M* Tris-HCl, 1 m*M* ethylenediamine tetraacetic acid (EDTA), pH 7.5.
10. Ribosomal RNA standard: diluted to 2 µg/mL in TE buffer.
11. Reverse transcriptase (RT) buffer.
12. Deoxynucleotide triphosphates (dNTPs).
13. Random hexamer primers.
14. Dithiothreitol (DTT).
15. Avian myoblastoma virus (AMV) RT (and buffer) or MultiScribe reverse transcriptase (and buffer), RNase-free H_2O.
16. TaqMan Master Mix (Applied Biosystems).
17. Forward and reverse primers.
18. Taqman probe.
19. ddH_2O.

3. Methods

There are several critical considerations to be taken into account when performing quantitative gene expression analysis, namely:

a. The isolation of high-quality, "clean" (i.e., DNA-free), undegraded RNA samples.
b. The precise quantitation of RNA template.
c. The careful control of experimental variables (e.g., pipetting errors).
d. The appropriate statistical data analysis and normalization of data.

The following sections provide methods that allow for the quantification of gene expression in a reliable manner that addresses all of these aspects. If followed carefully, these protocols should facilitate the generation of reproducible, high-quality data.

3.1. Quantitative Evaluation of G2A and GPR4 mRNA Expression in Cells and Tissue

Real-time TaqMan PCR can be applied to quantitate the expression of target mRNAs, such as GPCRs. This can be achieved across species using similar protocols with RNA samples isolated from a variety of sources, both those derived from cell culture experiments and those extracted from whole tissue. To exemplify these approaches, the present manuscript details the evaluation of two lysophospholipid receptors (G2A and GPR4) using RNA extracted from a variety of human cells in culture and in cardiac tissues isolated from the rat.

3.1.1. Cell Culture Growth Conditions and Treatments

The present report has focused upon mRNAs extracted from human leukocytes, e.g., monocytes, macrophages, platelets, T/B-lymphocytes, and so on (either derived as primary cells from several human donors or from established cell lines such as the THP-1 monocytic cell line). Cell culture conditions generally have little bearing on the RT-PCR process, and investigators should adhere to the suppliers recommendations for directions of growth conditions (although some cell types with intrinsically high "enzymatic" properties may require special care). What is of paramount importance, however, is that all efforts should focus on the rapid isolation of "nuclease-free" RNA samples (*see* **Subheading 4.1.**). Sterile cell culturing conditions will assist in this method, but care must be taken to avoid introducing extraneous contaminants at all points subsequent to the termination of the "in life" portion of any study. Because the protocols described herein use the principle of the PCR, nucleic acid levels can be determined in extremely small samples (this includes contaminating genomic DNA). Further, because TaqMan PCR relies on the generation of cDNA templates (by reverse transcription) from RNA samples, all reasonable attempts should be made to prevent ribonucleic acid degradation (RNase-free materials should be used, gloves should be worn, and so on). It is possible to expedite the expression analysis process by culturing cells in a 96-well plate format wherever possible (this is of particular utility under circumstances where multiple treatment conditions are being compared and contrasted).

3.1.2. Surgical Induction of Myocardial Infarction In the Rat

Issues relating to DNA contamination/nuclease contamination also apply to studies using materials harvested ex vivo or postmortem. In the present study, RNA has been extracted from adult male Lewis rat (300–400 g) hearts at various time points following coronary artery occlusion, an intervention that results in myocardial infarction and the subsequent development of left ventricular dysfunction.

Briefly, rats were anesthetized with isoflurane before surgery and myocardial ischemia was induced by permanent occlusion of the left anterior descending coronary artery. Following surgery, rats were sacrificed at various time points (0, 1, 4, 8, and 24 h and 3, 14, 28, and 56 d; n = 12 rats/time point). Hearts were divided into left and right ventricle for subsequent harvesting of RNA. In order to prevent RNA degradation, tissues were placed immediately into liquid N_2 and stored at –80°C until total RNA could be extracted. All protocols conformed to the Guide for the Care and Use of Laboratory Animals, US National Institutes of Health, NIH Publication No. 85–23.

3.2. Isolation of DNA-Free RNA

Removal of genomic DNA contamination from the RNA samples to be analyzed and avoidance of RNases is critical (*see* **Note 2**), especially in cases where PCR amplification utilizes TaqMan primers which span a single exon (because TaqMan PCR works most efficiently using small amplicons of approx 100 bp, low-fidelity genomic DNA amplification is less likely to become an issue if forward and reverse primers are separated by the large intronic stretches of DNA commonly found in genomic DNA). As discussed later, genomic DNA contamination can become an issue when studying GPCRs because of their genomic organization (most GPCRs are encoded as single exons).

In order to avoid the spurious amplification of genomic DNA from cDNA samples derived from total RNA, it is necessary to remove any deoxynucleic acids (i.e., genomic DNA) from the ribonucleic acid (i.e., total RNA) samples to be evaluated. DNA polymerase will not differentiate between templates derived from reverse transcription of total RNA (cDNA) or that present as a result of genomic DNA contamination.

3.2.1. Method 1 (DNase Treatment Protocol Prior to Reverse Transcription)

1. Add 1 µL DNase I to each 1 µg RNA sample to be processed (in 25–50 µL volume). This method uses Ambion DNase I and could be easily scaled up depending on the amount of RNA to be treated (RNase-free DNase must be used to avoid degradation of RNA samples prior to reverse transcription).
2. Mix samples well by flicking several times and spin down prior to incubation.
3. Incubate at 37°C for 10 min followed by a further incubation at 70°C for 5 min (this latter step is important because it heat-inactivates the DNase enzyme, a denaturation step that prevents subsequent damage of cDNAs generated during PCR process).

This standard DNase reaction can be scaled up by either changing the volume or the ratios (e.g., 10 µL of DNase I added to each 10 µg sample in 100 µL volume or 5 µL of DNase I to added to each 10 µg sample in 50 µL volume). The efficiency of the DNase procedure should be validated in a standard TaqMan assay using RNA samples that have not been subjected to a reverse transcription step (using a relatively abundant housekeeping gene as the target gene). RNA samples with threshold cycles (C_T) >40 in a standard 40-cycle PCR reaction can be considered "DNA free." For those who consider themselves novices to PCR, it is recommended that this "control" reaction be confirmed to ensure that "clean" RNA has been generated (because DNA polymerase cannot use RNA as a template, any amplicon generated from total RNA samples that have not been subjected to reverse transcription cannot be derived from mRNA and must result from the presence of unwanted [genomic] DNA contaminant). It is the recommendation of the authors that all investigators should do this. Alternatively, if DNA contamination of an RNA sample is suspected, this can be verified by gel electrophoresis. However, given the sensitivity of PCR approach, DNase-verification by TaqMan PCR is the preferred method (especially when dealing with GPCRs, which are typically expressed with low abundance). When an RNA sample is run out on a denaturing agarose gel, two discrete bands (18S and 28S RNA) should be observed. The presence of significant amounts of contaminating genomic DNA will manifest themselves as a "smear" within the gel upon staining with ethidium bromide.

If samples are to be processed immediately, DNase treatment of RNA samples can be incorporated into the RNA isolation process (*see* **Subheading 3.2.2., step 10**). The DNA-free RNA extraction protocol described below is simple to perform and provides RNA of a quality suitable for a number of purposes (cDNA synthesis/PCR, gene expression analysis, and so on). Many commercial kits are available for the extraction of RNA from cell extracts and tissues. Indeed, for cell-based applications, RNA can be extracted directly from 96-well plates in "reverse transcription"-ready form using a number of commercial kits such as RNeasy 96-well kit (Qiagen) or the RNAqueous MAG-96 kit (Ambion). In all other cases, the method outlined as follows will provide sufficient high quality RNA for downstream analysis. If it is not convenient to extract RNA from cells immediately, cells or tissue can be stored in RNAlater (Ambion/Qiagen) prior to RNA extraction for up to 12 mo post harvesting. Cell pellets are suspended in 100 µL phosphate-buffered saline (PBS) per 1×10^6 cells to which 5–10 vol RNAlater are added (for tissues, samples are stored in approx 10X vol of RNAlater). Samples can be stored at 25°C for 1 wk, 4°C for 1 mo and at –20°C indefinitely.

The following method can be used to purify RNA from most mammalian cells and tissue as well as the cells of lower eukaryotic and prokaryotic cells. This method utilizes an additional RNA purification step, making use of commercially available RNA binding columns generate ultrapure RNA for quantitative PCR analysis.

3.2.2. Method 2 (RNA Extraction Protocol Including a DNase Treatment Step)

1. Add 1 mL of Tri reagent per 5–10 × 10^6 cells or 10 cm^2 of culture plate. Homogenize lysate using a rotor–stator homogenizer.
2. Incubate samples for 5 min at 15–30°C.
3. Add 0.2 mL of chloroform per 1mL of Tri reagent and shake vigorously by hand for 20 s.
4. Allow the mixture to stand on ice for 2–10 min.
5. Centrifuge samples at 120*g* for 20 min at 4°C.
6. Remove the colorless upper aqueous phase, taking care not to contaminate it with the white DNA interphase or lower red protein organic phase. Transfer to a new 1.5 mL RNase-free tube.
7. Add 1 vol of 70% ethanol to the cleared supernate and mix immediately by pipetting or inversion. Do not centrifuge. A precipitate may form after the addition of ethanol, but this will not affect subsequent steps.
8. Apply up to 700 µL of the sample, including any precipitate that may have formed, to a RNeasy mini column placed in a 2 mL collection tube (supplied with kit). Close the tube gently, and centrifuge for 15 s at ≥100*g* (from this point, all centrifugation steps should be performed at 20–25°C). Re-apply the eluate, reusing the same collection tube. If the volume exceeds 700 µL, load aliquots successively onto the RNeasy column and centrifuge as above.
9. Pipet 350 µL Buffer RW1 into the RNeasy mini-column and centrifuge for 15 s at 100*g* to wash. Discard the flow-through
10. Add 10 µL DNase I stock solution to 70 µL Buffer RDD. Mix by gently inverting the tube. Buffer RDD is supplied with the RNase-Free DNase Set.
11. Pipette the DNase I incubation mix (80 µL) directly onto the RNeasy silica-gel membrane and place on the benchtop (20–30°C) for 15 min. Note that the DNase I incubation mix should be pipetted directly onto the RNeasy silica-gel membrane. DNase digestion will be incomplete if part of the mix sticks to the walls or the O-ring of the RNeasy column.
12. Reusing the collection tube in **step 9**, pipet 350 µL Buffer RW1 into the RNeasy mini-column and centrifuge for 15 s at 100*g*. Discard the flow-through.
13. Transfer the RNeasy column into a new 2 mL collection tube. Pipet 500 µL Buffer RPE onto the column. Centrifuge for 15 s at 100*g*. Discard the flow-through.
14. Add another 500 µL Buffer RPE to the RNeasy column. Close the tube gently and centrifuge for 2 min at 100*g* to dry the RNeasy silica-gel membrane.
15. Transfer the RNeasy column to a 1.5 mL collection tube (an Eppendorf tube can be used here). Pipet 33 µL RNase-free water directly onto the RNeasy silica-gel membrane. Close the tube gently, let it stand for 5–10 min at room temperature, and centrifuge for 1 min at 100*g* to elute. Re-apply eluate and spin again. Store sample at –20°C or proceed to quantification and reverse transcription.

For studies examining G2A and GPR4 expression in tissue specimens, total RNA was extracted from cardiac tissue (left and right ventricle) using similar protocols (Qiagen RNeasy Maxi kit) according to the manufacturers instructions (Qiagen, Inc., Santa Clarita, CA). When extracting RNA from whole organs such as rat hearts, tissues can be crudely cut into small pieces and are then powdered by hand using a mortar and pestle under liquid N_2. Once generated, powdered tissues are stored in Tri reagent (1 mL per 100–150 mg tissue). Alternately, tissues can be processed automatically in a high throughput mode using a Qiagen model MM300 Mixer Mill (not only is this method less tedious and slow, it is likely to provide more homogeneous RNA tissue extractions and reduces the sample-to-sample variability associated with manual extractions).

3.3. RNA Quantification

The accurate determination of total RNA concentrations is essential to the TaqMan process since all subsequent calculations are be based upon cDNA templates/PCR reactions derived from this starting material. Two methods are outlined below for the accurate quantification of RNA. The first is suitable for low throughput sample analysis (but is a less precise method) whereas the second is more amenable to simultaneous quantification of large sample collections (up to 86 high-quality RNA samples).

3.3.1. Method 1 (Low Throughput)

1. Generate a blank sample by pipetting 1 mL double-distilled water (ddH_2O) into a cuvet and measuring absorbence in a spectrophotometer at wavelengths of 260 nm (A_{260}) and 280 nm (A_{280}).
2. Dilute a small aliquot of RNA in ddH_2O to approx 2–10 μg/mL and transfer to a cuvet.
3. Measure the absorbence of the RNA sample at 260 nm and 280 nm.
4. Calculate RNA quantity using the following formula:

   RNA quantity (μg/mL) = $A_{260} \times 40 \times$ dilution factor.

In cases where only small amounts of RNA are being assessed, it may be appropriate to save the RNA after optical density (OD) determination. Protein contamination of RNA samples, which might reduce the efficiency of the reverse transcription reaction described below, can be measured by quantifying OD at a wavelength of 280 nm. Pure RNA will yield a 260/280 ratio of 1.8–2.0 whereas protein contamination will result in lower 260/280 ratio values (values <1.6 indicate serious contamination).

An alternative to that protocol described above, is one that is able to measure RNA (20 ng/mL–1 μg/mL concentration range) in a higher throughput mode (suitable for 96-well microplates). The DNA-free total RNA samples can be precisely quantitated using RiboGreen RNA quantitation reagent from Molecular Probes (note: the RiboGreen reagent also binds to DNA and, as such, the assay will not be "RNA-specific" *per se*).

3.3.2. Method 2 (High Throughput)

This protocol uses a 200 µL volume assay and measures RNA concentrations in the range 1–50 ng/mL in a 96-well TaqMan plate format. To prepare the RNA standards, dilute a 100 µg/mL stock to 2 µg/mL (50-fold) by taking 40 µL of RNA standard and adding to 1.96 mL of TE buffer (for "low range" standard curves, dilute the 2 µg/mL stock 20-fold into TE buffer to make 100 ng/mL stock).

1. Add 100 µL RNase-free TE buffer to wells A2–A7 and B2–B7 of the TaqMan 96-well plate.
2. Add 100 ng/mL RNA standard to wells A1 and B1 and to the contents of A2 and B2.
3. Make doubling dilutions by transferring the mixed contents of A2 to A3 and then A3 to A4 and so on, stopping at A6 (discard the excess 100 µL and leave A7 as TE buffer only to serve as a blank).
4. Repeat the same for B2. The dilution series of RNA (100 pg/µL down to 3pg/µL) are set in duplicate with 100 µL per well.
5. Prepare dilutions of the unknown samples you wish to quantitate in the remaining 82 wells (the sensitivity limit is approx 10 pg/µL RNA).
6. Dilute fluorescent RNA-binding probe 2000-fold in TE buffer (the probe is photolabile, so ensure that the reagent is protected from light).
7. Add 100 µL diluted probe to each well with multichannel pipet and mix.

Samples can then be read on ABI 7700 machine after 5 min at room temperature in the dark (ABI 7700 program commands are shown in uppercase in the following, but obviously it is advisable to read the manufacturer's instructions in detail before attempting to utilize the machine).

1. OPEN TaqMan, CLOSE default plate, FILE, NEW PLATE, select PLATE READ. (Also turn off ROX reference box in "Instrument... Advanced...").
2. Set up plate template with UNKNOWNS for samples and STANDARD for RNA standards, and use FAM presets (Ex 488 Em 530).
3. Put empty TaqMan plate in block and press PRE-PCR READ.
4. Put TaqMan plate containing samples into ABI7700 and press POST-PCR READ.
5. ANALYSIS, ANALYZE, then select ANALYSIS, DISPLAY, Rn.
6. Displayed will be the fluorescence values ready for standard curve and reading of samples.

3.4. cDNA Generation From RNA by Reverse Transcription

Two methods for generating cDNA from RNA are provided depending on whether small (approx 1 µg) or large (approx 10 µg) amounts of total RNA are being processed (*see* **Note 3**).

3.4.1. Method 1 (Low RNA Quantities)

This protocol uses AMV-RT to transcribe isolated RNA into cDNA:

1. Place 3 µL of each 0.33 µg/µL RNA sample (*see* **Notes**) in quadruplicate (three positive, one negative) into a MicroAmp Optical 96-well reaction plate. Heat to 55°C for 1 min and then immediately place on ice.
2. To each well add:

10 mM dNTPs	1 µL
250 ng/µL random hexamer primers	1 µL
ddH_2O	8 µL

3. Leave for 5 min at 25°C (note that for more dilute RNA samples, the amount of RNA can be increased and the amount of ddH_2O decreased to make a total volume of 13 µL).
4. To three of the quadruplicate sample wells add:
   4 µL 5X AMV-RT buffer
   2 µL of 0.1 *M* DTT
   1 µL AMV-RT
5. To the remaining sample (this "control" sample will lack RT and will act as the negative control for use in subsequent gene expression analysis) add the following:
   4 µL 5X AMV-RT buffer
   2 µL of 0.1 *M* DTT
   1 µL ddH_2O
6. Incubate all samples at 42°C for 1 h followed by 70°C for 15 min.
7. Dilute samples to 200 µL with DNase-free water and store at –20°C.

3.4.2. Method 2 (Large RNA Quantities)

As an alternative, High Capacity cDNA Archive kits (Applied Biosystems) can be used to reverse transcribe large quantities (up to 10 µg) of total RNA (100 µL reaction volumes). This is of utility where it is intended to run multiple TaqMan reactions (i.e., against hundreds of gene targets) and large quantities of cDNA template are required. Alternatively, the following basic protocol can be used:

1. Prepare a "master mix" for all samples (the maximum volume of RNA for 100 μL reaction is 70 μL, remember to add a 10% dead volume for experiments with more then one sample and keep all samples and reagents on ice):

10X RT buffer	10 μL
25X dNTP Mix	4 μL
10X Random Primers	10 μL
MultiScribe RT enzyme (50U/μL)	5 μL
Supernase (RNase inhibitor)	1 μL

2. Prepare all RNA samples by adding RNase-free H_2O up to 70 μL.
3. Add 30 μL of Master Mix to each sample.
4. Mix well by flicking several times.
5. Spin down briefly prior to incubation.
6. Incubate at 25°C for 10 min.
7. Incubate at 37°C for 2 h.

This reaction can be scaled down if using less then 10 μg total RNA (do not increase the volume above 100 μL as the efficiency of reaction will decrease significantly). The resultant cDNA samples can be stored cDNA in –80°C (but for data reproducibility, the number freeze–thaw cycles should be controlled and kept to a minimum as best possible).

3.5. TaqMan Real-Time PCR Primer and Probe Design

As is the case when attempting to ensure the fidelity of standard PCR amplification, selection of suitable TaqMan primers and probes is critical in the amplification of specific target cDNAs (i.e., avoiding co-amplification of genomic DNA, homologous cDNAs templates). In most cases, primers and probes are designed with the aid of software provided by thermal cycler manufacturers e.g., Primer Express software from Applied Biosystems (Foster City, CA). Although it is not always possible to adhere to all standard guidelines when it comes to designing primers (many GPCRs are encoded by a single intron and, therefore, the use of primers which span an exon is not possible in which case it is advisable to run "no RT" control reactions), it is suggested that several "rules" be adhered to as best is possible to try and ensure PCR fidelity:

1. Primer pairs should be specific for a target gene (promiscuity can be assessed using bioinformatics/Basic Local Alignment Search Tool (BLAST) searches to ensure that primers do not exhibit excessive homology with related "family" genes, pseudogenes, and so on, something that can be minimised by avoiding conserved "motifs").
2. Primers and probes should be designed towards regions of the target cDNA with a G:C content of 20 to 80% (PCR stringency is greater with primers designed against low G:C content cDNAs/amplicons).

3. Primers should be designed to generate relative small amplicons (approx 50–150 bp in size, primers should be designed to be as close to the probes as possible, the smaller the amplicon, the more efficient the amplification by *Taq* polymerase).
4. Probe and primer melting temperatures (T_m) should be 68–70°C and 58–60°C, respectively.
5. When designing probes, avoid "nucleotide repeats" (particularly >4 Gs).
6. Avoid placing Gs on the 5'-end of the probe.
7. Probes should be C > G-rich
8. For greatest fidelity, attempts should be made to limit the number of Gs and Cs in the last five nucleotides at the 3'-end of the primer to no more than two if possible (the presence of dA nucleotides ensures efficient catabolism of unwanted primer-dimers).

The following primers and probes were utilized for human G2A and GPR4:

human GPR4 forward primer	5'-AGG AGA TGG CCA ATG CCT C-3'
human GPR4 reverse primer	5'-TGG CTG TGC TCT TCC TCT TG-3'
human GPR4 probe	5'-CTC ACC CTG GAG ACC CCA CTC ACC T-3'
human G2A forward primer	5'-CCC GTA CCA CCT GGT TCT CC-3'
human G2A reverse primer	5'-CCG TGG ACA GGC ACA GAA AC-3'
human G2A probe	5'-TAC AGA GGA GAC AGG AAC GCC ATG TGC-3'

The following primers and probes were utilized for rat G2A and GPR4:

rat GPR4 forward primer	5'-ATG CTT CCC TCA CCC TGG AG-3'
rat GPR4 reverse primer	5'-GTG CAG GAT GAC AGT TGG GC-3'
rat GPR4 probe	5'-ACC ATT GAC CTC CAA GAG GAG CAC CAC-3'
rat G2A forward primer	5'-ATG CAA CAG GAA ATG CCA CG-3'
rat G2A reverse primer	5'-CTG TAC ACC GCC ACC AGG AC-3'
rat G2A probe	5'-ATG AGC TGC CAC ACG TTG TCC TAC GAG-3'

3.6. TaqMan Real-Time PCR Conditions

Real-time PCR is performed using an ABI Prism® Model 7700 or 7900HT sequence detection system (or some similar CCD detector/thermal cycler apparatus). TaqMan RT-PCR reaction kits are available and come provided with detailed, easy-to-use protocols. The following is an overview of such protocols:

3.6.1. Method 1 (cDNA Standards)

1. The first step of the PCR protocol requires the generation of a standard curve. The precise details of the standard curve used are dependent upon the properties of the Taqman primers and probes that have been designed (discussed previously). For Taqman primers that will not amplify from genomic DNA (i.e., primers that span and intron-exon boundary) the standard curve is made from cDNA generated from cells that express the target tissue of interest (discussed later). For TaqMan primers/probes that are able to amplify from genomic DNA (as is the case with most GPCRs which are usually encoded by a single exon), this can be used as the substrate for standard curve generation. Concentrations for both standard curves are as follows:
2. Prepare serial dilutions of cDNA template (*see* **Note 4**)

   1.6 μg/μL 400 ng/μL 100 ng/μL 25 ng/μL 6.3 ng/μL 1.6 ng/μL 0.4 ng/μL H_2O
3. Prepare a TaqMan reaction master mix. For each reaction well add the following (remembering to add a 10% dead volume if making up reagent for more than one well):

   12.5 μL 2X Master Mix
   1.5 μL 5 μ*M* forward primer
   1.5 μL 5 μ*M* reverse primer
   0.5 μL 5 μ*M* probe
   4.0 μL ddH_2O

4. Using a multichannel pipette (*see* **Note 4**), transfer 5 μL cDNA (5ng/μL) from the cDNA stock plate into a new MicroAmp 96-well reaction plate (depending on RNA yield, water can be substituted for additional cDNA and *vice versa*).
5. Using a repeating pipet, add 20 μL Master Mix into each well.
6. Seal the plate using MicroAmp 96-well optical lids (PE Biosystems) and centrifuge at 15*g* for 1 min.
7. Place the plate in the ABI 7700 or 7900HT Prism Sequence Detector. Typical reaction conditions are as follows (but are influenced by specific probe/primer designs):

   50°C for 2 min
   95°C for 10 min
   then 40 cycles of 95°C for 15 s and 60°C for 1 min

Following completion of the reaction, the plate can be discarded (or retained to check for product generation/amplicon size, and so on, using electrophoresis on a 4% agarose gel).

3.6.2. Method 2 (Genomic DNA Standards)

If TaqMan primer sets are designed within a single exon, relative quantification can also be accomplished using samples of known concentrations of genomic DNA. Based on the following calculations, the quantity of each sample is assessed by extrapolating a standard curve from the range of diluted standards with known genomic DNA (allowing for the expression of data as copies of mRNA of interest detected in specific amount of the RNA sample). Taking into consideration the stability of the diluted standards described below, genomic standards can be prepared every 2–3 wk.

Diploid genome contains:

6×10^9 base pairs (bp)
$6 \times 10^9 \times$ 660 (molecular weight) is equivalent to 4×10^{12} g (for diploid)
2×10^{12}g/L constitutes 1 mole
1 mole approximates to 6.02×10^{23} copies DNA (for haploid)
this is equivalent to 12.04×10^{23} single strand (ss) copies of template
1 g/L approximates to 6.02×10^{11} ss copies/L or 6.02×10^5 ss copies/µL

1. Calculate number of copies/µL in the concentrated genomic DNA sample ("?"µg/µL = "?" $\times 6 \times 10^5$ ss copies/µL).
2. Then dilute with ddH_2O to 0.2×10^5 ss copies/µL i.e., 5 µL equals 1×10^5 copies suitable for 20, 25, and 50 µL reaction volumes (dilute accordingly for smaller reaction volumes).
3. Prepare 1:10 dilutions for 10,000, 1000, 100, 10, and 1 copies, respectively.
4. Diluted standards can be stored in 4°C.

3.7 Setting Up TaqMan Reactions

Unlike the 7700 model, real-time PCR reactions can be set in two high-density formats (96- and 384-well) using the ABI 7900HT Sequence Detection System. The recommended template concentrations per reaction in the 96- and 384-well formats is equivalent to 25–50 ng and 10–20 ng of total RNA used for reverse transcription, respectively (the recommended volumes are 25–50 µL and 10–20 µL in 96- and 384-well format, respectively). It is recommended that, where possible, each sample be run at least in duplicate. Single data point per sample can be acceptable in the experiments with large number of experimental replicates ($n \geq 3$).

1. Dilute cDNA samples to a final concentration of 5 ng/µL (based on RNA amount taken to the reverse transcription).
2. Precisely pipet 5 µL template into the wells (robotic liquid handling can be used in this step for more high throughput approach).
3. Accurately pipet genomic standards (each dilution in duplicate).
4. Prepare TaqMan Master Mix (remember to add 10% dead volume). Reagents for a 25 µL reaction are:

2X TaqMan Master Mix	12.5 µL
100 µ*M* Forward Primer	0.22 µL (900 n*M* working concentration)
100 µ*M* Reverse Primer	0.22 µL (900 n*M* working concentration)
10 µ*M* TaqMan Probe	0.25 µL (100 n*M* working concentration)
ddH_2O	6.81 µL

5. Add 20 μL of Master Mix to each well using a multidispenser pipet.
6. Seal the plate using optical adhesive cover (Applied Biosystems) and centrifuge 15*g* for 1 min (this will remove air bubbles which might interfere with the optics of the SDS machine).
7. Place the plate in the SDS system (e.g., Model 7700 or 7900HT) and create plate document by specifying experimental samples, standards, and appropriate controls (e.g., no template control).
8. Select appropriate detector type and for most conditions use the universal PCR conditions recommended by the vendor:

   50°C for 2 min and 95°C for 10 min followed by
   40 cycles of 95°C for 15 s and 60°C for 1 min

3.8. Data Analysis

As outlined in **Subheading 1.4.**, Taqman ABI Prism (e.g., Models 7700/7900) SDS measure emitted fluorescence in each of 96- or 384-individual reaction wells during both the denaturation and annealing/extension phases of the PCR reaction. Detailed descriptions of the data analysis are provided by the manufacturer.

Briefly, emission/fluoresence data captured by the system's CCD camera is used to construct amplification plots for each well (sample wells and those used for standard curve construction) using the PE Biosystems software provided. Using the amplification curves generated, the user is able to extract threshold cycle values (C_T) of emission for each well coupled to a quantity value (generated by the inclusion of a standard curve on each Taqman plate). These values can be expressed as ng DNA, copy number, or normalized to the average log of one or more housekeeping genes of choice, e.g., in the in vivo studies described herein, expression was determined relative to that observed on d 0 of the study after values were normalized to two the expression of two housekeeping genes, namely cyclophilin and rpL32 (known to provide more consistent expression that GAPDH in these tissues).

Before calculating copy numbers for any given gene of interest, "baseline" and "threshold" values must be set for the SDS instrument. For most applications, the vendor-suggested baseline 3–15 cycles can be used. Defining this "threshold" is critical (the threshold cycle, C_T, occurs where the system software begins to detect the increase in signal associated with exponential growth of PCR product). During the exponential phase of PCR, the amount of product is proportional to the initial copy number of the template. This relationship changes as the rate of amplification approaches a plateau (hence thresholds should be set within the exponential phase of PCR). Once the threshold is defined, the C_T (threshold cycle) values and copy numbers (if standards are present on the plate) can be exported for statistical data analysis. As noted above, to assess RNA quality, template loading efficiency and cDNA conver-

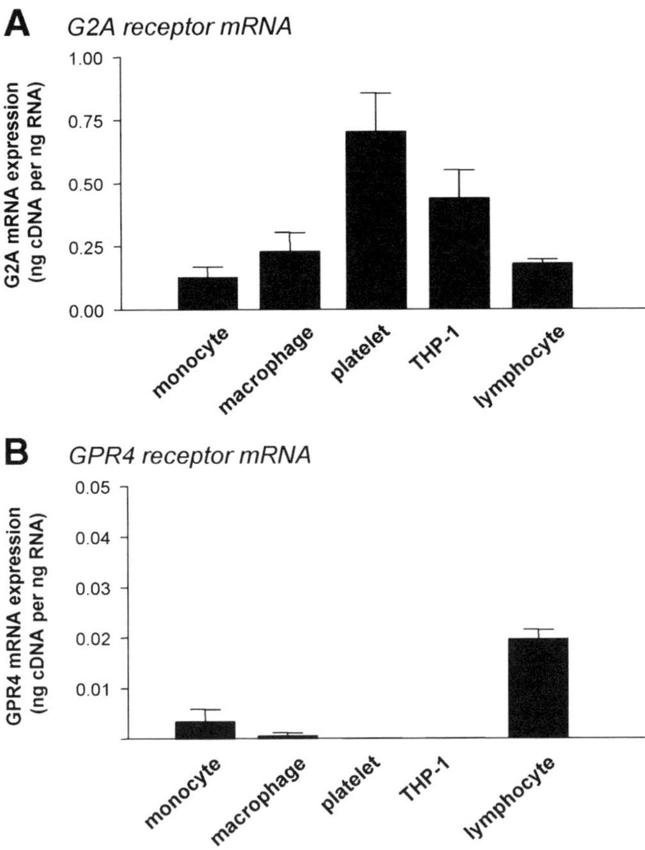

Fig. 2. Quantitation of relative (**A**) G2A and (**B**) GPR4 mRNA expression in human leukocytes in culture as determined by TaqMan® real-time polymerase chain reaction.

sion efficiency expression profiles for more then one housekeeping gene should be generated. These data can be then used for normalization.

Adjusting for multiple comparisons, a one-way analysis of variation (ANOVA) followed by a Dunnett's test is run separately for each gene for comparisons relative to d 0 values (for calculating confidence intervals). Changes are considered statistically significant where $p \leq 0.05$.

3.9. Quantitative Expression of G2A and GPR4 mRNA in Cells and Tissues as Estimated Using TaqMan Real-Time PCR

Quantitative analysis of mRNA expression in human leukocytes revealed significant GPR4 expression in monocytes (both fresh primary human monocytes and the human monocyte cell line THP-1; **Fig. 2A**) and in human mac-

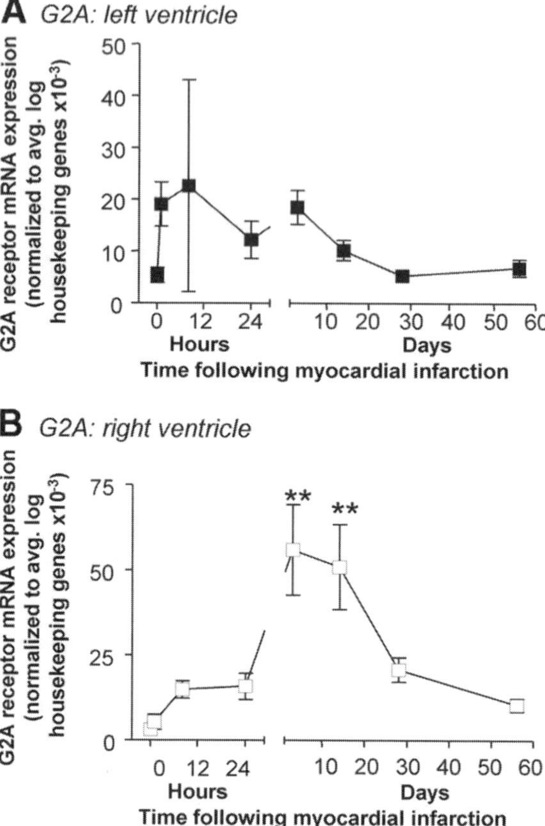

Fig. 3. Quantitation by TaqMan® real-time PCR of the relative temporal changes in G2A mRNA expression in (**A**) the infarcted left and (**B**) noninfarcted right ventricle of the rat following permanent coronary artery ligation.

rophages (albeit at significantly lower levels of expression). In contrast, GPR4 mRNA transcription was not detected in either human platelets or T/B-lymphocytes. This profile differed somewhat from that observed with G2A where receptor mRNA was detected at approx 10–20-fold higher levels (**Fig. 2B**) than that recorded with GPR4 and where expression was evident in all cells tested (in particular, in platelets and lymphocytes).

G2A mRNA expression was highly upregulated approx 30-fold in the right ventricle following the induction of a myocardial infarct in the rat (peak transcript expression observed 1–2 wk post injury was statistically significant; $p < 0.01$; **Fig 3A**). Expression was also enhanced in the infarcted left ventricle compared to the contralateral (noninfarcted) although this failed to reach statistical significance (**Fig 3B**). In contrast to G2A, GPR4 transcript levels were

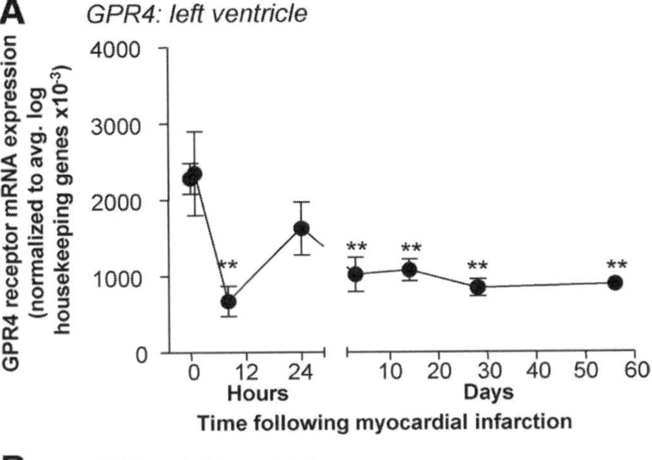

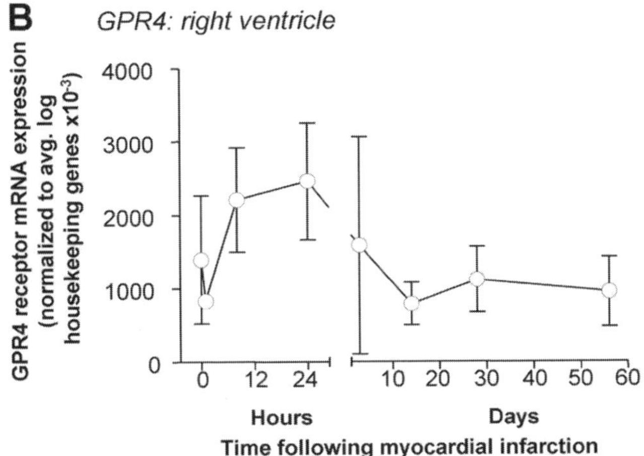

Fig. 4. Quantitation by TaqMan® real-time polymerase chain reaction of the relative temporal changes in GPR4 mRNA expression in (**A**) the infarcted left and (**B**) noninfarcted right ventricle of the rat following permanent coronary artery ligation.

actually found to be significantly attenuated (by approx 50%) in the right ventricle essentially for the entire 8 wk study period following left ventricular infarction (with the exception of 1 h and 24 h, $p < 0.01$; **Fig 4A**). Further, in contrast to G2A, GPR4 mRNA expression was not significantly altered (less than twofold) in the infarcted left ventricle (less than twofold changes which were not statistically significant; **Fig 4B**).

Because changes appeared to be most profound with G2A rather than GPR4, the present data might suggest that the latter lysophospholipid receptor plays as more significant role in the etiology of heart failure (based, at least, on the data generated in this rodent infarction model). Clearly, such supposition requires

OF BRISTOL
LIBRARY
MEDICAL

significant additional work because transcriptional changes are not always directly correlated with changes in protein translation, and so on (although it is also of note that the in vitro data suggests that G2A is capable of modulating the functions of both myeloid [monocytes, macrophages] and lymphoid [T-cells]) cells. Nevertheless, although this preliminary "target validation" information is far from conclusive, it does perhaps allow one to prioritize one target over the other; i.e., perhaps G2A is more worthy of immediate inspection if one is interested in delineating the role of LPC and SPC in vascular inflammation following infarction and the subsequent progression of heart failure.

4. Notes

1. A basic understanding of rudimentary molecular biological techniques and laboratory practices is advantageous. Numerous publications are available to serve as introductory guides to working with nucleic acids, PCR, and so on. It is also advisable to familiarize oneself with some of the basic practical tenets which underpin the discipline of molecular biology (access to a reference book describing basic molecular biology protocols is clearly advantageous ***[15]***). Some basic guidelines to follow would be:
   a. Avoid contaminating RNA samples, i.e., wear clean gloves, lab coats, and so on, articles that can be readily changed when contamination is suspected (this practice will also limit the introduction of unwanted nucleases), and work in an area of the laboratory and with tools and reagents that are dedicated for PCR use.
   b. Use "core mixes" where ever possible (but exercise caution to prevent "cross-contaminating" samples, e.g., avoid the "repeated use" of pipet tips [preferably use the aerosol-resistant type of tip]).
2. As with any PCR-based methodology, the most important consideration in the preparation and analysis of RNA is to inhibit rapidly and efficiently the endogenous ribonucleases present in virtually all living cells. Failure to do so results in the generation of poorly degraded RNA and leads to inconsistency when attempting to quantitate expression differences between treatments, samples, and so on. Many procedures for the isolation and analysis of RNA exist:
   a. When isolating RNA, work quickly and ensure that all tubes, solutions/reagents and pipet tips used are RNase free.
   b. RNA is subject to degradation under conditions of excessive shear stress, so exercise caution when drawing RNA-containing fluids into pipets (because large shearing forces can be generated in pipet tips with small openings).
3. In order to provide standardization across samples, all RNA samples of interest should be reverse-transcribed into cDNA templates at the same time. Note also that, although the RNA quantity utilized in the protocol is an optimal amount for the reverse transcription reaction, the method will efficiently reverse transcribe down to 50 ng/μL and up to 1μg/μL RNA (or 10 μg/μL total RNA).

4. It is imperative that volumes be measured accurately (any error here in the amount of starting cDNA template; for example, will literally become amplified during the PCR process). As with most molecular biology techniques, the use of homogeneous "core mixes" is advisable (discussed previously) wherever possible to avoid any differences in sample-to-sample reagent variability. Care should be taken to avoid cross-contamination of samples through the use of common pipet tips, and so on.

References

1. Douglas, S. A., Ohlstein, E. H., and Johns, D. G. (2004) Cardiovascular pharmacology and drug discovery in the 21st century. *Trends Pharmacol. Sci.* **25**, 225–233.
2. Bleicher, K. H., Bohm, H. J., Muller, K., and Alanine, A. I. (2003). Hit and lead generation: beyond high-throughput screening. *Nat. Rev. Drug Disc.* **2,** 369–378.
3. Robas, N., O'Reilly, M., Katugampola, S., and Fidock, M. (2003). Maximizing serendipity: strategies for identifying ligands for orphan G-protein-coupled receptors. *Curr. Opin. Pharmacol.* **3,** 121–126.
4. Cacace, A., Banks, M., Spicer, T., Civoli, F., and Watson, J. (2003) An ultra-HTS process for the identification of small molecule modulators of orphan G-protein-coupled receptors. *Drug Disc. Today* **8,** 785–792.
5. Zhu, K., Baudhuin, L.M., Hong, G., (2001) Sphingosylphosphorylcholine and lysophosphatidylcholine are ligands for the G protein-coupled receptor GPR4. *J. Biol. Chem.*, **276,** 41,325–41,335.
6. Kabarowski, J. H., Zhu, K., Le, L. Q., Witte, O. N., and Xu, Y. (2001) Lysophosphatidylcholine as a ligand for the immunoregulatory receptor G2A. *Science,* **293,** 702–705.
7. Gräler, M. H. and Goetzl, E. J. (2002) Lysophospholipids and their G-protein-coupled receptors in inflammation and immunity. *Biochim. Biophys. Acta* **1582,** 168–174.
8. Xu, Y. (2002). Sphingosylphosphorylcholine and lysophosphatidylcholine: G-protein-coupled receptors and receptor-mediated signal transduction. *Biochim. Biophys. Acta,* **1582,** 81–88.
9. Ross, R. (1999). Atherosclerosis: an inflammatory disease. *New Engl. J. Med.*, **340,** 115–126.
10. Frangogiannis, N.G., Smith, C.W., and Entman, M.L. (2002). The inflammatory response in myocardial infarction. *Cardiovasc. Res.*, **53,** 31–47.
11. Wilson, S., Bergsma, D.J., Chambers, J.K., (1998). Orphan G-protein-coupled receptors: the next generation of drug targets? *Br. J. Pharmacol.*, **125,** 1387–1392.
12. Lindsay, M.A. (2003). Target discovery. *Nature Rev.*, **2,** 831–838.
13. Holland, P.M., Abramson, R.D., Watson, R., and Gelfand, D.H. (1991). Detection of specific polymerase chain reaction product by utilizing the 5'-3' exonuclease activity of Thermus aquaticus DNA polymerase. *Proc. Natl Acad. Sci. USA* **88,** 7276–7280.
14. Heid, C.A., Stevens, J., Livak, K. J., and Williams, P.M. (1996). Real time quantitative PCR. *Genome Res.,* **6,** 986–994.
15. Sambrook, J., Fritsch, E.F., and Maniatis, T. (eds.) (1989) *Molecular Cloning: A Laboratory Manual.* Cold Spring Harbor Laboratory. Cold Spring Harbor, NY.

4

mRNA

Detection by In Situ *and Northern Hybridization*

Alessandra P. Princivalle, Rachel M. C. Parker, Terri J. Dover, and Nicholas M. Barnes

1. Introduction

The ability to detect mRNA by either *in situ* hybridization histochemistry (ISHH), first described in 1969 by Gall and Pardue and John et al. ***(1,2)*** or Northern hybridization, first described by Alwine et al. ***(3)***, has become a very powerful technique in many research areas, including that of receptor research. The applications of these techniques are many and include (1) direct assessment of the presence, distribution, and modulation under different physiological conditions of specific RNA species ***(4,5)***; (2) molecular investigations of potential mRNA splice variants and region-specific heterogeneity in multimeric-receptor subunit potential expression ***(6,7)***; (3) indirect detection of receptor-expression to support the existence of the receptor when highly-selective ligands (*see* Chapter 5) or antibodies (*see* Chapter 8) are unavailable for receptor localization studies ***(8)***; and (4) investigation of molecular changes in pathological states and the possible modes of action of drugs used to treat such conditions ***(9–11)***. Changes at the molecular level to alter mRNA expression represent rapid changes within a cell; therefore, it can be envisaged that such studies on human biopsy and post mortem tissue will lead to an array of important diagnostic tools. Furthermore, the combination of ISHH and immunohistochemistry (*see* Chapter 8) offers a powerful strategy to study the

From: *Methods in Molecular Biology, vol. 306: Receptor Binding Techniques: Second Edition*
Edited by: A. P. Davenport © Humana Press Inc., Totowa, NJ

co-existence of mRNA and the translated polypeptide product ***(12)***, with consistent results from the two approaches allowing greater confidence to be attached to the significance of the findings. Alternatively, the co-localization of one mRNA species with a peptide/protein phenotypically characteristic of a certain cell type allows the putative function of the protein to be proposed ***(13,14)*** which subsequently focuses further investigation.

Both ISHH and Northern hybridization exploit the principle that single-stranded nucleic acid sequences anneal to their complementary nucleic acid sequence. It therefore follows that when the single-stranded nucleic acid sequence is tagged or labeled (to produce a probe), the location of this hybridization can be detected. Northern hybridization is a relatively rapid method of detecting the presence, abundance, and size of specific RNA species within the population of cells from a given region: the RNA is first extracted from its tissue source, size fractionated by electrophoresis, transferred from the electrophoresis gel, and immobilized onto a membrane phase before being hybridized with a labeled complementary nucleic acid probe ***(15,16)***. In comparison, ISHH allows a specific RNA species to be detected directly at its site of expression, revealing its cellular localization and relative abundance ***(15,16)***. Both methods basically comprise the following seven steps:

1. *Probe labeling.* The choice of probe and label depends on the requirements of the research being undertaken (*see* **Tables 1** and **2**). (The cloning techniques needed to produce suitable vector templates for cDNA and riboprobe synthesis will not be covered in this chapter.)
2. *RNA isolation.* To minimize RNA degradation and maximize signal detection, it is critical that RNase-free conditions are maintained and tissue is collected, stored, and fixed correctly.
3. *Prehybridization tissue treatment.* The sensitivity of the method may be increased in several ways at this stage by employing one or a number of steps, depending on the nature of the tissue source, to (a) maintain RNA integrity (e.g., tissue fixation and use of RNase inactivators); (b) help reduce nonspecific background labeling by various treatments (e.g., delipidation, acetylation, and pre-incubation with hybridization solution prior to the addition of the labeled nucleic acid probe); and (c) in the case of ISHH, improve probe access by proteinase K treatment and, for paraffin-embedded tissue sections, removal of the wax.
4. *Hybridization.* Optimal temperature for hybridization is, as a general rule, at 20–25°C below the melting point temperature (Tm) of the nucleic acid probe ***(16–18)***, where the Tm for:

DNA/DNA = $81.5 + 16.6\log[Na^+] - 0.62(\%\ \text{formamide}) + 41(G{+}C) - 500/(\text{probe length})$

RNA/RNA = $79.8 + 18.5\log[Na^+] - 0.35(\%\ \text{formamide}) + 58(G{+}C) + 12(G{+}C)^2 - 820/(\text{probe length})$

DNA/RNA = mean of Tm (DNA/DNA) and Tm (RNA/RNA).

Table 1
Choice of Probe

Probe	Advantages to use	Disadvantages to use	Labeling method	Reference
cDNA (200–500 bases)	Easy to generate; stable, long, high specific activity	Need molecular biology knowledge, less efficient hybridization than RNA probes, presence of both DNA strands may decrease sensitivity	PCR, nick translation, or random primer method	***27,28***
Oligomers (20–50 bases)	Convenient; no molecular biology knowledge necessary; antisense and sense strands easily synthesized and labeled, stable; short, so good tissue access; high specificity	Need knowledge on designing a suitable sequence and access to a synthesizer, relatively insensitive, may need a cocktail of a number of oligomers to increase sensitivity	T4 polynucleotide kinase, terminal deoxynucleotidyl transferase	***29–32***
Riboprobes (200–500 bases)	Very sensitive (good for detecting less abundant mRNA species), good signal-to-noise ratio, RNase treatment after hybridization allows further background reduction, probe is strand-specific, generate both sense and antisense probes with high specific activity from the same vector.	Requires subcloning into a suitable promoter vector, difficult to generate, and requires special protection against RNase degradation	In vitro transcription, using T3-,T7-, or SP6-specific RNA polymerase	***33***

PCR, polymerase chain reaction.

Table 2
Choice of Label

Label	Advantages to use	Disadvantages to use	Reference
Nonradioactive (e.g., fluorescein, digoxigenin, biotin)	No special safety precautions required, quick results (within several days), labeled probe stable for up to 1 yr, provides cellular resolution	Nonquantitative, less sensitive than radioactive detection, tissue permeability is very critical, endogenous biotin may hinder accurate detection	***30, 31, 34***
Radioactive (e.g., [^{3}H], [^{35}S], [^{33}P], [^{32}P])	Sensitive, cellular resolution obtained with low-energy emitting isotopes (e.g., [^{3}H] and [^{35}S]) for *in situ*, quick results obtained with high-energy emitting isotopes (e.g., [^{32}P]), but with loss of resolution, therefore useful for Northern hybridization, semiquantitative	As a result of the half life, the labeled probe must be made fresh, special safety precautions are required, low-energy emitters need long exposure times and are slow to yield results, high-energy emitters scatter signal and only provide low resolution	***16, 30***

It therefore follows that the Tm value is contingent on a number of factors, including probe length and nucleotide base base composition (GC pairs have a greater influence on overall duplex stability as a result of their use of three hydrogen bonds, whereas AT pairs only utilize two hydrogen bonds). Other factors which will influence the optimal temperature for hybridization include homology of the probe to the target nucleic acid sequence (i.e., the number of mismatches between the sequence of the probe and the target nucleic acid; this may be apparent if the mRNA from a different animal species is being detected. However, for short oligonecleotide probes under optimal conditions, to maximize the signal-to-background detection one or two mismatches may prevent the detection of the RNA under study) and the concentrations of the salts and denaturing agent (e.g., formamide) used in the hybridization buffer.

5. *Posthybridization washing.* This step is designed to remove nonspecific background caused by any unbound and loosely bound probe, which may be present after the hybridization step as a result of weak homology with related RNA species or nonspecific interactions with other cellular components. Posthybridization wash stringency is directly proportional to temperature (where the most stringent wash is approx at 10–15°C below the Tm value) and inversely proportional to the salt concentration. Thus, the signal-to-background ratio can be further optimized by manipulation of these two parameters.
6. *Signal detection.* Photographic emulsion, film, or nonradioactive detection methods are available, depending on the nature of the tag or label used to visualize the probe (*see* **Table 2**).
7. *Controls.* It is vital to include positive and negative controls within a series of experiments to test the ability to obtain consistent, sensitive, and specific detection of the desired RNA species.

This chapter describes two highly sensitive, selective, and reproducible methods, successfully employed in our laboratory, for analyzing mRNA expression. The first half details methods for ISHH. These studies used radioactively labeled oligonucleotide probes to identify prodynorphin (PPD) mRNA within rat brain (*see* **Fig. 1**) and spinal cord *(4)*. This probe was used to evaluate the modulation of a molecular marker to, in turn, assess the involvement of neurokinin receptors in models of long-term hyperalgesia *(4)*. A further study, also utilizing a radiolabeled oligonucleotide probe, identified mRNA encoding the $GABA_{B1a}$ protein (*19*; **Fig. 2**) to assess alterations in the expression of this transcript in resected hippocampi from patients with temporal lobe epilepsy. The second method employs radioactively labeled 750 base antisense and sense riboprobes, corresponding to amino acid residues 62 to 312 of a human 5-HT_3 receptor cDNA sequence (h5-HT_3R riboprobe; 20) to identify specific 5-HT_3 receptor mRNA within the human central nervous system (**Fig. 3**). The distribution of 5-HT_3 receptor-binding sites has been characterized in the human brain, where they are suggested to be involved in many important physiological roles, such as in memory and learning, anxiety, and emesis *(21)*. Therefore,

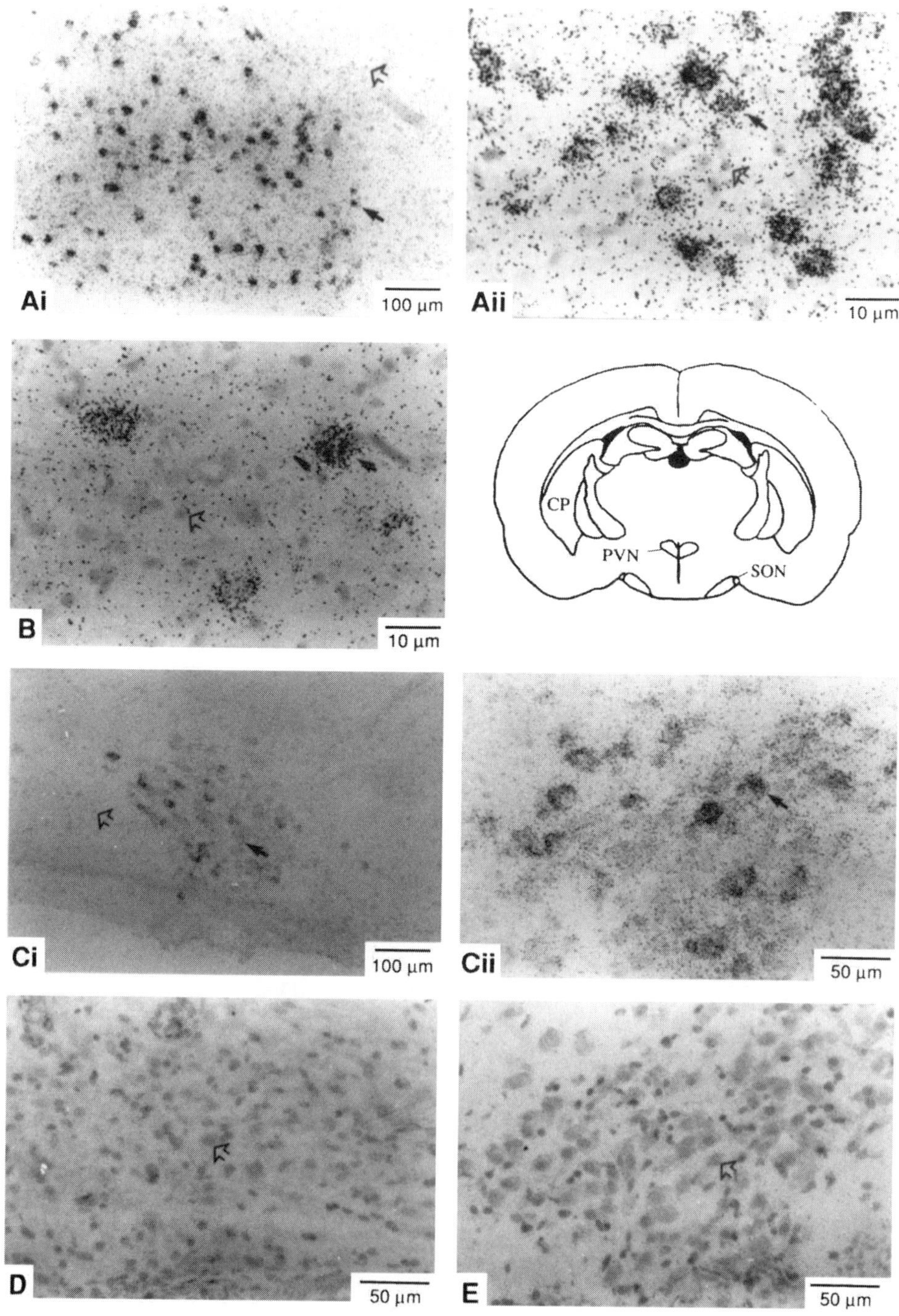
Ai
100 μm
Aii
10 μm
B
10 μm
CP
PVN
SON
Ci
100 μm
Cii
50 μm
D
50 μm
E
50 μm

it is relevant to phenotypically determine which subpopulation of neurones express this receptor in order to elucidate a role for this receptor subtype in such processes. The second half of the chapter, from **Subheading 3.2.**, describes the application of Northern hybridization, which was employed to determine the selectivity of the various 5-HT_3R riboprobes used in our ISHH studies and to locate 5-HT_3 receptor mRNA within different nervous system regions of human and rat (*see* **Fig. 3**). These protocols should provide a framework from which to work and adapt to other applications.

In addition, although not covered in the present chapter, the use of microarray array is particularly suitable to identify mRNAs and allows monitoring the expression of thousands of genes in a single experiment *(22)*. Comparable to *in situ* hybridization and Northern blotting, base-pairing (i.e., A-T and G-C for DNA; A-U and G-C for RNA) or hybridization, is the underlying principle. The technique provides a medium for matching known and unknown DNA samples based on simple base-pairing rules, and with automation, the high throughput of numerous samples is achievable.

2. Materials

2.1. Chemicals and Solutions

It is important to take precautions against RNase contamination when making the following solutions (*see* **Note 1**).

Fig. 1. *(continued from facing page)* Preprodynorphin mRNA expression in untreated rat brain, using *in situ* hybridization histochemistry. Low- and high-power light field photographs showing *in situ* hybridisation histochemical identification of prodynorphin (PPD) mRNA expression, restricted to discrete areas within the brains of normal, untreated rats (*see* schematic diagram of a rat brain section). Positively labeled neurones, as shown by a dense aggregation of silver grains around their nuclei (filled arrows), are present in **(A)** the caudate putamen (CP) as seen at low power (**Ai**) and higher power (**Aii**); **(B)** the paraventricular nuclei (PVN); and **(C)** two examples of labeling in the supraoptic nuclei (SON) at low power (**Ci**) and higher power (**Cii**). Note, in comparison, the low and evenly distributed background density of silver grains (approx 10 grains per 10 μm^2) overlying nonexpressing nuclei (open arrows) and the cytoplasm of these regions and also in nonexpressing areas, such as surrounding the CP (**Ai**) and SON (**Ci**). Adjacent sections pretreated with RNaseA (1 µg/µL) for 60 min at 37°C, before hybridization with the PPD oligonucleotide probe or hybridized with a similar concentration of the complementary sense probe to the PPD oligonucleotide display a similar level of even background with no evidence of clustering characteristic of labelled cells, as shown for the SON region in **D** and **E**, respectively.

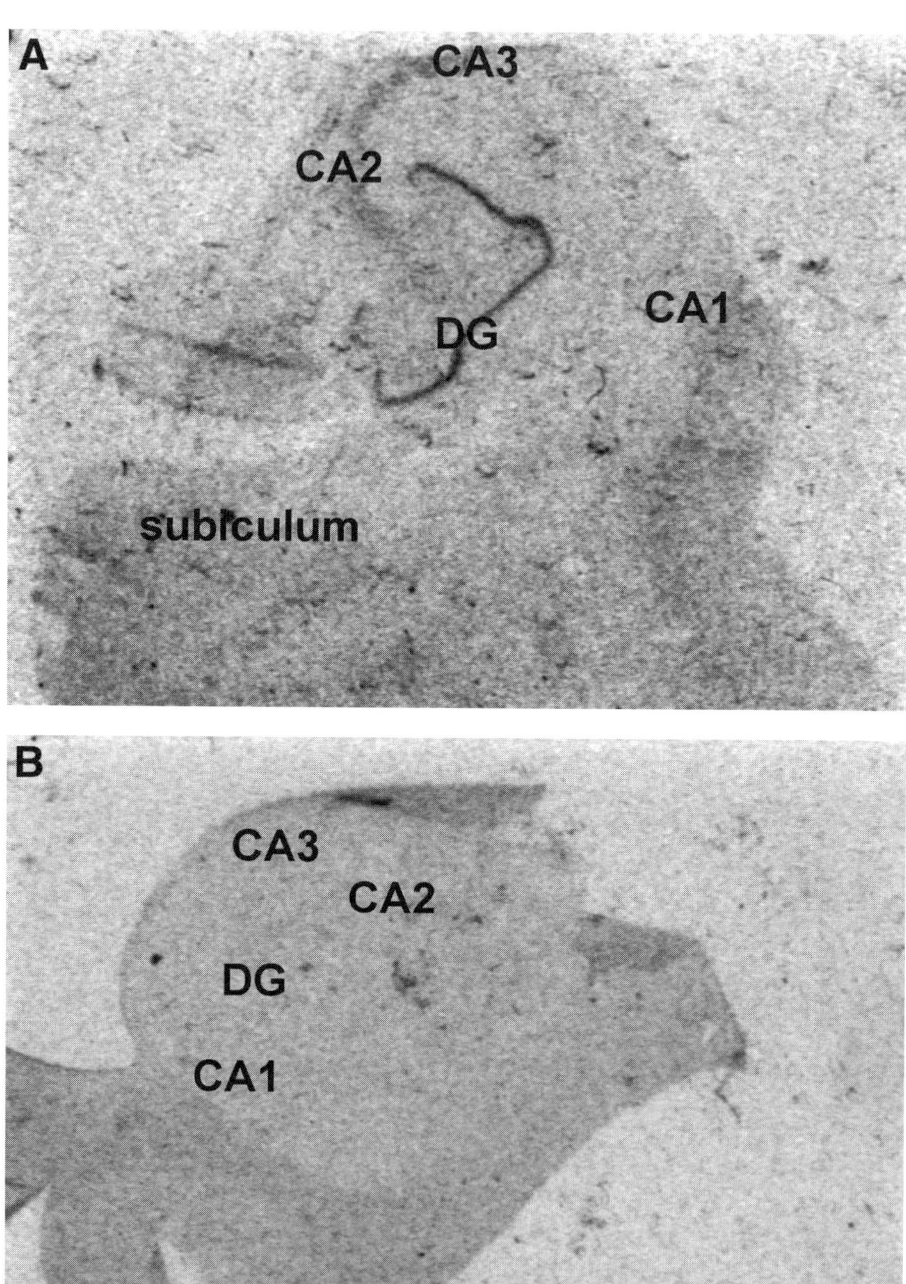
A
CA3
CA2
DG
CA1
subiculum
B
CA3
CA2
DG
CA1

1. 2-mercaptoethanol.
2. [^{32}P]-αUTP: (Uridine 5'-triphosphate-[α^{32}P], triethylammonium salt; specific activity >3000 Ci/mmol, 10 mCi/mL).
3. [^{35}S]-αdATP: (deoxyadenosine 5'-(α-thio) triphosphate-[^{35}S], triethylammonium salt (*in situ* grade); specific activity >1250 Ci/mmol, 12.5mCi/mL).
4. [^{35}S]-αUTP: (Uridine 5'-(α-thio) thiotriphosphate-[^{35}S], triethylammonium salt (SP6/T7 grade); specific activity >1250 Ci/mmol, 40mCi/mL).
5. Absolute ethanol.
6. Acetone.
7. Acid alcohol: add two to three drops of concentrated HCl to 70% ethanol solution.
8. Agarose: e.g., Ultrapure
9. Agarose (1%)/formaldehyde solution: for 100 mL, dissolve 1 g agarose in 72.1 mL diethylpyrocarbonate (DEPC)-treated dH_2O, cool to approx 60°C, then add 10 mL of 10X 3-[*N*-Morpholino]propanesulfonic acid (MOPS) running buffer. In a fume hood, add 17.7 mL of 37% solution of formaldehyde, to give a final concentration of 2.2M. Allow to cool before pouring the gel (agarose sets at approx 45°C).
10. Alkaline H_2O: add one drop of concentrated ammonia solution to 300 mL dH_2O, make fresh.
11. Chloroform.
12. Chloroform: Isoamylalcohol 49:1 mixture: mix in a fume cupboard, store at 4°C wrapped in foil.
13. Chromic acid: dissolve 10% (w/v) potassium dichromate in autoclaved dH_2O, very slowly and carefully add 10% (v/v) concentrated sulfuric acid, and mix with a glass rod. Handle this solution with care and store it at room temperature in a glass container with a tight-fitting lid, clearly labeled hazardous and corrosive. This solution can be used several times.
14. Cold sterilization solutions: (1) 3% H_2O_2 in DEPC-treated dH_2O, (2) 70% ethanol in DEPC-treated dH_2O, (3) 0.1N NaOH containing 1 m*M* ethylenediamine tetraacetic acid (EDTA) in DEPC-treated dH_2O.
15. Concentrated ammonia solution.
16. Concentrated HCl.
17. Decon.

Fig. 2. *(continued from facing page)* $GABA_{B1a}$ mRNA expression in human hippocampus, using *in situ* hybridization histochemistry. Autoradiographic images of *in situ* hybridization using [^{35}S]-labeled oligonucleotides to $GABA_{B1a}$ isoform mRNA in hippocampus from post mortem control (**A**). Adjacent sections pretreated with RNaseA (1 µg/µL) for 60 min at 42°C, before hybridization with the $GABA_{B1a}$ oligonucleotide probe or hybridized with a similar concentration of the complementary sense probe to the $GABA_{B1a}$ oligonucleotide display a similar level of even background with no evidence of clustering characteristic of labeled cells (**B**). Scale bar 2 mm.

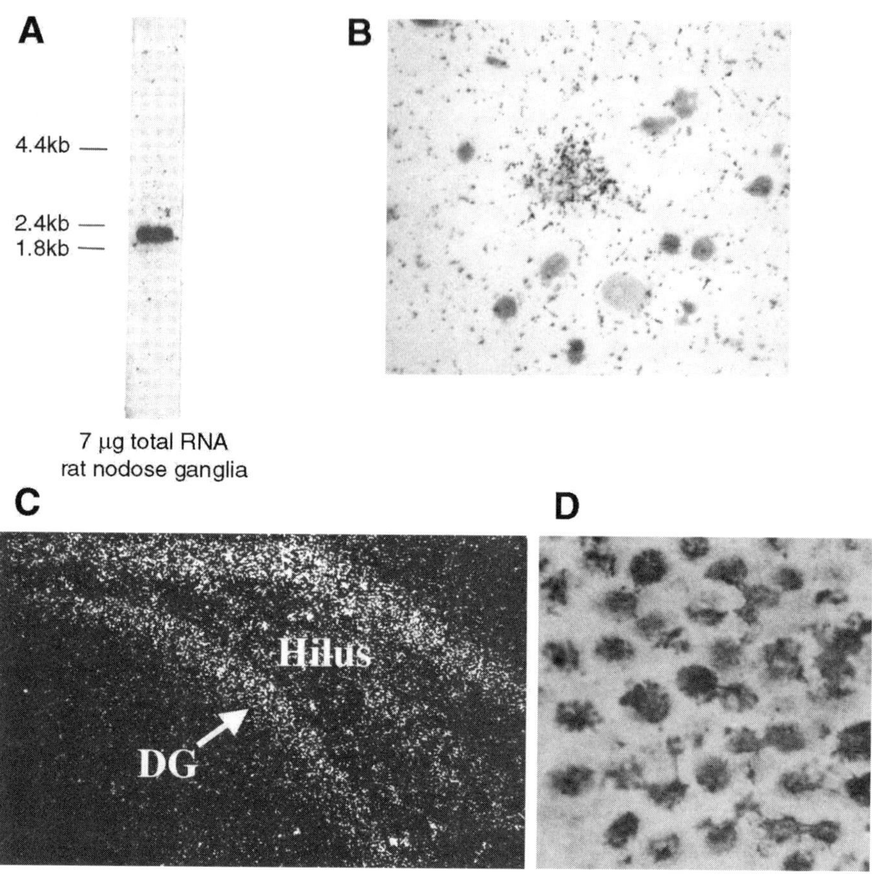

Fig. 3. Use of riboprobes to detect mRNA encoding the human 5-HT_{3A} receptor subunit (**A–B**) or rat NK-1 receptor (**C–D**). (**A**) The selectivity of 5-HT_3 receptor subunit antisense riboprobe was tested by Northern hybridization. A Northern blot of total RNA extracted from rat tissue known to highly express a 5-HT_3 receptor subunit, against a [^{32}P]-mouse 5-HT_3 receptor subunit antisense riboprobe clearly shows a single band of molecular weight corresponding to rat 5-HT_3 receptor mRNA ***(20)***, using the protocol detailed in this chapter. The position of molecular weight markers are shown on the left. This result was also obtained when RNA extracted from a mouse neuroblastoma cell line, highly expressing native murine 5-HT_3 receptors, was probed with [^{32}P]-mouse and [^{32}P]-human 5-HT_3 receptor subunit antisense riboprobes, but no bands were detected when identical samples were probed with the equivalent sense riboprobes (data not shown). This shows the ability of a single riboprobe to detect mRNA from different species. (**B**) Use of the [^{35}S]-human 5-HT_3 receptor subunit antisense riboprobe to label, with cellular resolution, 5-HT_3 mRNA in human hippocampus (hilus; brightfield). The labeled cell (heavy level of silver grains overlaying histologically stained cells) is a large, presumably GABAergic neurone. Note the lack of radiolabel associated with small cells in this region *(see* **ref.** ***26***). (**C–D**) Use of

18. De-ionized formamide: mix 100 mL formamide per 10 g mixed-bed ion exchange resin 20–50 mesh (BioRad AG 501-X8) for 30–60 min at room temperature, in a fume cupboard, filter twice through Whatman no. 1 filter paper, dispense into aliquots, store at –20°C.
19. Denhardt's solution (50X): 1% (w/v) Ficoll type 400 (e.g., Sigma F2637), 1% (wt/vol) PVP e.g., Sigma P5288), 1% (w/v) bovine serum albumin (BSA; fraction V; e.g., Sigma A7030), made up in DEPC-treated dH_2O. Filter through a millipore 0.2-μm filter, store in 10 mL aliquots at –20°C.
20. DePeX mounting medium.
21. Photographic developer.
22. DEPC-treated dH_2O (0.05%): in a fume hood, add 0.5 mL DEPC (stock stored at 4°C; very toxic) per liter dH_2O, shake well, then stand for 20 min; shake again and leave standing at room temperature for >2 h but <24 h, then autoclave for 35–40 min to destroy excess DEPC. All of the solutions are made up with DEPC-treated water, unless otherwise stated.
23. Dithiothreitol (DTT; 1 *M*): make a 1 *M* solution of DTT in DEPC-treated dH_2O, store at –20°C in autoclaved microcentrifuge tubes (this solution goes off very quickly at room temperature).
24. Dry ice.
25. EDTA (0.5 *M*): add EDTA to DEPC-treated dH_2O, stir with a magnetic flea, dissolve by adjusting to pH 8.0 with NaOH (10 *M*), then autoclave.
26. Embedding medium.
27. Eosin Y (1%) in 80% ethanol: make 80% ethanol solution with dH_2O, dissolve 1% Eosin Y in this, filter the solution twice through Whatman No. 1 filter paper before use.

Fig.3 *(conitnued from facing page)* radioactive [^{35}S]-labeled (C; darkfield) and digoxygenin-labeled riboprobe (**D**) to detect mRNA encoding the neurokinin-1 receptor (NK-1) receptor in the rat hippocampus (**C**; dentate gyrus [DG; hippocampal subregion] surrounding the hilus; **D**, cellular resolution of individual dentate granule neurones). The riboprobes were synthesised using the polymerase promoters T7 (sense) or SP6 (antisense) by in vitro translation incorporating [^{35}S]- (**C**) or digoxygenin-tagged UTP ([**D**]) RNA labeling kit; Boeringher Manheim, Germany; following manufactured instructions). The riboprobes were used to label the NK-1 mRNA in an identical protocol as the labeling of 5-HT_3 mRNA described in this chapter. Once hybridized, digoxygenin molecules incorporated within the riboprobes were visualized, with cellular resolution, by standard immunohistochemical techniques (antidigoxygenin primary antibody, which was subsequently labeled using a biotin-conjugated secondary antibody. The primary/secondary antibody complex was then visualized by addition of avidin-conjugated alkaline phosphatase and the subsequent incubation with the alkaline phosphatase substrate, BCIP/NPT (5-bromo-4-chloro-3-indolylphosphate)/nitoblue-tetrazolium salt), which generates a blue-purple reaction product.

28. Ethanol solutions containing 0.3 *M* ammonium acetate: add freshly made 3 *M* ammonium acetate solution (dissolved in DEPC-treated dH_2O) to 50%, 70%, 80%, and 90% ethanol solutions made up in DEPC-treated dH_2O, to give a final concentration of ethanol solutions containing 0.3 *M* ammonium acetate.
29. Ethanol solutions for histological staining: make 95%, 80%, and 70% ethanol solutions with dH_2O, store at room temperature.
30. Ethanolamine solution: 0.15 *M* NaCl, 0.1 *M* triethanolamine, 0.25% (v/v) acetic anhydride. Make fresh.
31. Ethidium Bromide solution (EtBr): make a 10 mg/mL solution of EtBr in dH_2O; store at 4°C in a sealed, dark bottle protected against light (very hazardous/carcinogenic).
32. Photographic fixer.
33. Formaldehyde: 37% solution (12.3 *M*); hazardous; store in chemical box.
34. GTC denaturing solution: at 65°C, dissolve 4 *M* guanidine thiocyanite (GTC; e.g., Sigma G9277) (disrupts membranes and inhibits RNases) in a solution containing 25 m*M* sodium citrate solution (pH 7.0), 0.5% *N*-lauroylsarcosine (disrupts nucleoprotein complexes) (prepare a 10% stock solution and store at 4°C; NB may need to heat stock solution if it has precipitated), and 5 m*M* EDTA (pH 8.0) (RNase inhibitor); once dissolved, add 0.8% (w/v) 2-mercaptoethanol (RNase inhibitor), filter the final solution into an autoclaved bottle with a 0.2-µm millipore filter, and store at 4°C for up to 3 mo.
35. HCl (0.2 *N*).
36. Histoclear.
37. Hybridization solution (2X) for ISHH: dissolve 20% (w/v) dextran sulfate (helps probe anneal) in DEPC-treated dH_2O, heated to 60°C, then add the following ingredients (final concentration); 1.2 *M* NaCl, 20 m*M* Tris buffer, pH 7.5, 2X Ficoll type 400 (0.04% w/v), 2X PVP (0.04% w/v), 10X BSA (0.2% w/v) (96–99% albumin; store at 4°C), 2 m*M* EDTA (to chelate nuclease activating ions), 0.02% (200 µg/mL) salmon sperm sonicated single-strand (ss)DNA (store at –20°C), 0.001% (10 µg/mL) glycogen (stock stored at –20°C), and 0.01% (100 µg/mL) type X-SA baker's yeast tRNA (stock stored at –20°C) (the last three stabilize the probe and are nonspecific hybridization blockers). Mix the final solution well, taking care not to denature the BSA (NB solutions containing BSA cannot be autoclaved). Store 0.5 mL aliquots in autoclaved microcentrifuge tubes at –20°C for up to 6 mo. When ready to use, thaw an aliquot and add an equal volume (i.e., 500 µL) of deionized formamide and mix well to give 1X hybridization solution containing 50% (v/v) de-ionized formamide. The final solution should be pH 7.0–7.5.
38. Isoamyl-alcohol (3 methyl-1 butanol).
39. Isopentane (2-methyl butane): e.g., BDH 103616V.
40. Isopropanol (2-propanol).
41. Linearized DNA template: approx 0.8 µg; store at –20°C (*see* **Note 7**).
42. Liquid nitrogen.
43. Mayer's Hematoxylin solution: filter twice through Whatman no. 1 filter paper before use.

44. MOPS running buffer (10X): for 1 L; dissolve 4.1 g sodium acetate in 700 mL DEPC-treated dH_2O, add 41.9 g MOPS, pH to 7.0 with 10 *M* NaOH, then add 10 mL of 0.5 *M* EDTA (pH 8.0), make up volume with DEPC-treated dH_2O, autoclave then store at 4°C.
45. NaOH (0.05 *N*): make with DEPC-treated dH_2O.
46. Nuclear emulsion: e.g., Ilford K5 (www.ilford.com); store at 4°C in total darkness; lasts approx 2 mo; 20 mL diluted emulsion is just sufficient for 35 slides.
47. Oligonucleotide: of known concentration (can be stable for years at –70°C) (*see* **Note 3**).
48. Paraformaldehyde (4% w/v) in 0.1 *M* phosphate-buffered saline (PBS): For 1 L, warm 500 mL DEPC-treated dH_2O to 65°C in a fume cupboard, slowly add 40 *g* paraformaldehyde. Add approx 200 µL NaOH (10 *M*) and vigorously stir to facilitate dissolving of paraformaldehyde (takes <1 h). Cool, add 500 mL 0.2 *M* PBS, pH to 7.5; this will keep 1 mo at 4°C, after which it may polymerize (it will also polymerize if autoclaved); this is a very hazardous chemical.
49. Phenol saturated with TE buffer: Phenol is very hazardous; store at 4°C, always use in a fume hood, and be extremely careful not to inhale fumes or expose to skin.
50. PBS; 0.1 *M*: 10 m*M* Na_2HPO_4, 3 m*M* KCl, 1.8 m*M* KH_2PO_4, 140 m*M* NaCl, pH to 7.4, and autoclave. Will keep at 4°C for 2–3 wk.
51. Poly-L-lysine subbing solution: mix 100 mg poly-L-lysine (mol wt >300,000, store at –20°C) in 500 mL DEPC-treated dH_2O in an autoclaved bottle (this is enough to generate approx 500 slides); make poly-L-lysine solution fresh.
52. RNA Loading Buffer: 0.4% bromophenol blue , 0.4% xylene cyanol, 1 m*M* EDTA (pH 8.0), 50% glycerol. Store stock at 4°C. To minimize contaminations, aliquot 1 mL volumes into autoclaved microcentrifuge tubes.
53. RNA markers: 0.24–9.5 Kb Ladder, store at –20°C.
54. RNA polymerases: supplied with 5X transcription buffer and 100 m*M* DTT; SP6 (e.g., Promega P1085), T7 (e.g., Promega P2075), or T3 (e.g., Promega P2083); store –20°C or –70°C for long-term storage.
55. RNase-Free DNase: e.g., Promega M6101; store at –20°C.
56. RNaseA solution buffer: 10 m*M* Tris-HCl (pH 7.6), 0.5 *M* NaCl, 1 m*M* EDTA, pH to 8.0, and autoclave; will keep at 4°C.
57. RNaseA solution: dissolve ribonuclease A (RNase A; bovine pancreas; stock stored –20°C) in autoclaved dH_2O (which has not been DPEC-treated). Store aliquots at –20°C at a concentration of 12.5 mg/500 µL. When ready to use, boil aliquot for 1 min to remove DNase activity then dilute this aliquot in RNaseA solution buffer to give a final concentration of 25 µg RNase/mL buffer.
58. rRNasin: Promega N2511; 20–40 units/µL; store at –20°C.
59. Salmon sperm sonicated ssDNA: e.g., Sigma D9156; stored at –20°C; 10 mg/mL. Immediately before use, denature this solution at 100°C for 5 min, then rapidly cool on ice.
60. Scintillation fluid.
61. Sodium dodecyl sulfate (SDS), 10% solution.
62. Sephadex slurry: equilibrate Sephadex G25 powder in 1X TE buffer (pH 8.0) overnight, store slurry at 4°C.
63. Silica gel: 6–20 mesh.

64. Sodium acetate solution (2 *M*): dissolve sodium acetate in DEPC-treated dH_2O, adjust to pH 4.0 with glacial acetic acid, autoclave, store solution at 4°C.
65. Sodium acetate solution (3 *M*): dissolve sodium acetate in DEPC-treated dH_2O, adjust to pH 5.0 with glacial acetic acid, autoclave, store solution at 4°C.
66. SSPE (20X): 3 *M* NaCl, 0.2 *M* NaH_2PO_4, 20 m*M* EDTA.
67. Standard sodium citrate solution (SSC; 20X): 3 *M* sodium chloride, 0.3 *M* tri-sodium citrate in DEPC-treated dH_2O, adjust to pH 7.0, and autoclave, will keep 2–3 wk at room temperature.
68. T4 DNA polymerase: e.g., Sigma D0410; 10 U/µL; store at –80°C.
69. TdT enzyme: Terminal deoxynucleotidyl transferase supplied with 5X tailing buffer; e.g., GIBCO BRL; 18008-011; store at –20°C.
70. TE buffer (1X): 10 mM Tris-HCl, 1 m*M* EDTA, pH 8.0 with 10 *M* NaOH, and autoclave.
71. Tris (1 *M*): add 6.35 g Trizma Hydrochloride to 1.18 g Trizma base in 50 mL of autoclaved dH_2O, then autoclave to give a solution of pH 7.5 at 25°C. We do not use DEPC-treated dH_2O in case there is any residual DEPC present, to react with Tris-HCL.
72. Unlabeled nucleotide mix: add 10 µL of each stock of 10 m*M* ATP, 10 m*M* GTP, and 10 m*M* CTP (store at –80°C) to 10 µL of DEPC-treated dH_2O to give a final concentration of 2.5 m*M* of each base. Store the mix in aliquots of 2 µL at –20°C.
73. Xylene.

2.2. Equipment

1. Adequate radiation protection: perspex shielding, correct decontamination procedures and methods of disposing of radioactive waste.
2. Film cassettes and autoradiography film, e.g., Kodak BioMax (Amersham Biosciences, http://www.amershambioscinces.com).
3. Gel apparatus and power pack.
4. Glass microfiber filter discs: e.g., Whatman 1822 025 (http://www.whatman.com).
5. Hand-held beta counter.
6. Homogenator or sonicator.
7. Hybond-N+ membrane: e.g., RPN 203B (Amersham Biosciences, http://www.amershambioscinces.com).
8. Hybridization vials and mesh: (e.g., Hybaid, www.thermo.com).
9. Image analysis system.
10. Intensifying Screens: optional.
11. Light box.
12. Microscope.
13. Microscope coverslips, slide mailers and slides.
14. Slide storage boxes: sealable (e.g., Kartell). Boxes which hold 25 slides are useful, as these will comfortably hold silica gel along with 20 slides (a managable number to assay at one time).
15. Ultraviolet (UV) spectrometer and 1 cm^2 quartz cuvet.

16. Ultraviolet (UV) transilluminator.
17. Whatman chromatography paper (3 MM).

3. Methods

3.1. In Situ *Hybridization Histochemistry*

It is important to take precautions against RNase contamination when carrying out the following protocols (*see* **Note 1**).

3.1.1. Probe Labeling

3.1.1.1. Spun Column Preparation

1. Plug a sterile 1 mL syringe barrel with glass microfibre filter paper, packing it in place with the syringe plunger.
2. Slowly pipet 1 mL of Sephadex slurry into the column, taking care to avoid air bubbles.
3. Sit the column in a screw-capped microcentrifuge tube and spin this set up in a 15-mL disposable centrifuge tube at >1100*g* for 4 min, allowing the buffer to run through into the collecting microcentrifuge tube at the bottom.
4. Repeat the pipetting and spinning (**steps 2** and **3**) until the column contents stabilize at 1 mL of Sephadex G50.
5. Add a known volume (e.g., 100 µL) of 1X TE buffer to the top of the Sephadex bed, place the column in a fresh microcentrifuge tube, and repeat **step 3**. The volume recovered should be equal to that added to the top of the column.
6. Repeat **step 5** twice more.
7. Add 100 µL of 1X TE buffer to the top of the Sephadex bed, seal the column with Parafilm to prevent evaporation, and store upright at 4°C until ready to use (*see* **Note 2**).

3.1.1.2. 3'End-Labeling of Oligonucleotide Probes With [^{35}S]

1. Set up the following reaction in autoclaved microcentrifuge in order:

		final
Diluted $GABA_{B1a}$ probe (*see* **Note 3**)	2.0 µL (approx 150ng)	approx 50 n*M*
5X Tailing buffer (supplied with the enzyme)	2.5 µL	1X
[^{35}S]-a-dATP (*see* **Notes 4** and **5**)	12.0 µL (approx 150 µCi)	approx 1.4 µ*M*

2. Equilibrate the above reaction mix to 37°C (takes approx 5 min)
3. Add TdT enzyme (15 U/µL) 3.0 µL — 1.5 unit/µL
   autoclaved DEPC-treated dH_2O — 5.5 µL (to make up volume to 25 µL)

   total volume — 25.0 µL

4. Flick to gently mix, then microcentrifuge the reagents down for 2–3 s to remove air bubbles which may inhibit TdT enzyme activity.
5. Incubate the mixture for 2.5 h at 42°C. (Meanwhile, let the spun column [as prepared in **Subheading 3.1.1.1.**] equilibrate to room temperature for approx 30 min).
6. Stop the reaction by cooling the microcentrifuge tube on ice.
7. Prespin the spun column for 4 min at 1100*g* and discard the run-through buffer.
8. Place the spun column in a fresh collecting microcentrifuge tube.
9. In duplicate, spot a 1 μL aliquot of prespun reaction mix (to measure total cpm) onto a 0.5 cm^2 of filter paper in a scintillation vial, add scintillation fluid, vortex, and count using a suitable program on a β-counter.
10. Apply the rest of sample (approx 60 μL) to the center of the top of the spun column Sephadex bed and spin the column for 4 min at 1100*g*. The unincorporated label will remain in the column while the probe runs through. The column can now be correctly disposed of as radioactive waste (*see* **Note 4**).
11. Retain the labeled probe in the collecting tube, measure the final volume (this should be of equal volume to that put on the column) and take duplicate 1-μL aliquots for scintillation counting (as in **step 9**) to obtain a measure of incorporated cpm/μL (dpm/μL can be calculated if the counting efficiency of the β-counter is known; it is usually approx 95% for [^{35}S] isotopes). We on average obtain at least 1–1.5 × 10^6 dpm/μL (i.e., 1 × 10^8 dpm per [^{35}S]-oligonucleotide labeling reaction).
12. The labeled probe can be stored at –20°C if it is to be used within the next few days. Alternatively, it can be kept up to 2 wk at –70°C, after which the level of disintegration necessitates repurification through a fresh Sephadex spun column. We usually label the oligonucleotide 1 d to use within 24 h.
13. From the scintillation readings, calculate:

(a) $$\% \text{ incorporation} = \frac{\text{incorporated cpm (postspin counts; \textbf{step 11})} \times 100}{\text{total cpm (prespin counts; \textbf{step 9})}}$$

the % incorporation should average at least 70% and should not be below 50%

(b) Specific activity [cpm (or dpm]/mol and cpm [or dpm]/μg) assuming 100% of the probe is labeling and recovered (we on average obtain 2.2–4.4 × 10^{18} dpm/mol, i.e., >10^9 dpm/μg).

3.1.1.3. Riboprobe Labeling With [^{35}S] by In Vitro Transcription

1. In an autoclaved microcentrifuge tube, at room temperature to avoid DNA precipitation, add in order:

		final
5X transcription buffer	2 μL	1X
100 m*M* DTT (*see* **Note 6**)	1 μL	10 m*M*
rRNasin (20–40 U/μL)	1 μL	2–4 U/μL
Linearized h5-HT_3R cDNA template (*see* **Note 7**)	1 μL	0.8 μg /reaction

Unlabeled nucleotide mix (*see* **Note 8**)	2 μL	500 μ*M* of each base
^{35}S-αUTP (100 μCi) (*see* **Notes 4** and **9**)	2.5 μL	approx 10 μ*M*
RNA polymerase (20 U/μL) (*see* **Note 10**)	0.5 μL	1 U/μL
	10μL	

2. Flick to gently mix, then microcentrifuge the reagents down for 2–3 s to remove air bubbles which may inhibit enzyme activity.
3. Incubate at 37°C for 2 h (meanwhile, let the spun column [as prepared in **Subheading 3.1.1.1.**] equilibrate to room temperature for approx 30 min).
4. After the 2 h incubation, add 1 U of RNase-free DNase and incubate at 37°C for 15 min to destroy the DNA template (need 1 U DNase/μg DNA).
5. Add 5 μL of 100 mm DTT stock and make the volume up to 80 μL with DEPC-treated dh$_2$o (*see* **Note 6**).
6. Purify the riboprobe through a spun column as in **Subheading 3.1.1.2., Steps 7–12.**
7. From the scintillation counter readings, calculate:
   (a) cpm/μL (or dpm/μL, if the counting efficiency is known; it is usually approx 70% for [^{32}P] and 95% for [^{35}S]), (for [^{35}S] expect approx 10^7 dpm/μL) and the total cpm (or dpm), knowing the total volume (which should be equal to the volume added to the column)
   (b) % incorporation = incorporated cpm (postspin counts; **Subheading 3.1.3., step 11** × 100 / total cpm (prespin counts; **Subheading 3.1.3., step 9**)
   (should obtain a value >40%, <90%)
   (c) total RNA made = % incorporation × theoretical maximum yield (assuming one-fourth of the total nucleotides are labeled UTP)

e.g., to calculate the theoretical maximum RNA yield, using 100 μCi [^{35}S]-αUTP:

the maximum UTP incorporation	= 100 μCi/the specific activity of the radiolabel
	= 100 μCi/1200 nmol UTP
therefore maximum base incorporation	= 4 × 100/1200 nmol bases
molecular weight of RNA	approx 330 ng/nmol
therefore, the maximum yield of RNA	approx 4 × 100/1200 × 330 ng

   (d) specific activity = total cpm/ total RNA made, as calculated from the values obtained in **steps a** and **c**) (this should be >10^9 cpm/μg RNA for [^{32}P]- and [^{35}S]-aUTP).

8. Check the quality and molecular weight of the riboprobe on a RNA denaturing gel: set up a denaturing gel as detailed in **Subheading 3.2.2.3., step 1–10**, except load 0.5 μL riboprobe (or at least 5 × 10^5 cpm) in 4.5 μL loading solution with 1 μL of loading dye per well, and run the gel along with appropriate RNA size markers. Dry the gel (with the markers removed) on a gel drier, wrap in plastic wrap and expose to autoradiography film at –70°C (as for membrane blot described in **Subheading 3.6.1.**) for up to 2 d. Develop the film as detailed in **Subheading 3.2.6.2**. and analyze the film with reference to the

markers, as described in **Suheading 3.2.8.** A good quality probe will produce few bands of the expected length, indicating successful in vitro transcription. Poor quality probes will yield a smear of smaller sized products as a result of degradation or poor quality template cDNA, or give bands of higher molecular weight than expected, indicating incomplete linearization of the template.

3.1.2. RNA Collection

3.1.2.1. Preparation of Clean, RNase-Free Microscope Slides

1. Taking precautions as set out in **Note 1**, stack slides in glass slide racks and wash them overnight in a sealed plastic container containing chromic acid (taking care, as the acid bath is very hazardous).
2. Rinse the slides in tap water then wash in running water overnight.
3. Wash the slides in 2% warm decon for 30 min before washing them in running water for 12 h to 1 d.
4. Soak the slides for 15 min in DEPC-treated dH_2O and repeat with fresh DEPC-treated dH_2O.
5. Soak the slides in absolute ethanol for 10 to 15 min to remove any remaining grease.
6. Dry the slides at 37°C, then immediately proceed to Subheading **3.1.2.2.**

3.1.2.2. Subbing Microscope Slides

1. Dip the clean, RNase-free slides for 3 min each in 0.2 *N* HCl, followed by DEPC-treated dH_2O and finally in acetone.
2. Dry the slides at 50–60°C for 15 min.
3. Dip the slides in freshly prepared poly-L-lysine subbing solution for 10 s, remove the slides, and then repeat dipping in the same solution for another 10 s.
4. Rinse the slides in DEPC-treated dH_2O for 10 min (this decreases static and dust attraction).
5. Dry slides overnight at 50–60°C, then store, sealed in dust-free slide boxes, at room temperature (the slides will keep for 6–8 mo; *see* **Note 11**).

3.1.2.3. Tissue Collection and Section Cutting

1. Immediately after death, isolate the tissue of interest, using sterile instruments and working as quickly as possible to reduce RNase activity and RNA degradation.
2. Mount the tissue in a minimal amount of embedding medium on a cryostat chuck (excess embedding medium can be trimmed off with a scalpel blade prior to cutting).
3. Rapidly immerse the mounted sample in a beaker of isopentane, cooled to –45°C in a dry ice bath and freeze rat spinal cord (1–1.5 cm long segments) for 3 mins and whole rat brain for 5 min. The temperature of the isopentane is critical; any lower than –45°C may cause the tissue to fracture when cut; any higher, and the sample may not be rapidly frozen.

4. Wrap the frozen tissue in Parafilm, place in a sealed water-tight container, and store at –70°C.
5. Transfer the fresh frozen tissue from storage to the cryostat on dry ice, allowing the tissue to slowly equilibrate to the cryostat temperature (takes >30 min). We find a chamber temperature of –20°C to –25°C works best for rat spinal cord and –16°C to –19°C for rat brain (the optimal temperature will depend on the size and type of the tissue and on the cryostat).
6. Trim tissue until intact, undamaged sections are obtained (*see* **Note 12**). Collect individual 10μm thick sections by thaw-mounting onto the poly-L-lysine subbed slides (*see* **Subheading 3.2.2.**), which have been kept at room temperature (we usually mount 2–3 brain sections or 5–10 spinal cord sections per slide).
7. Periodically take sections for histological staining (*see* **Subheading 3.6.3.**) to check section quality.
8. Allow the sections to dry at room temperature for several minutes, then return them to the cryostat, until they can be transferred on dry ice to the –70°C freezer, for storage in sealed slide boxes containing silica gel desiccant to prevent frost building up on the slides. Sections should be used in ISHH as soon as possible.

3.1.3. *Prehybridization* (see **Table 3**)

1. Sterilize all prehybridization containers and hybridization boxes (*see* **Note 1**).
2. Bring boxes containing the slides to room temperature (takes approx 10 min) before opening them, to prevent condensation forming on the sections.
3. Stack the slides to be used in a slide rack. Return the remaining slides to the –70°C as soon as possible. We find that 25–50 slides (enough to fill one to two slide racks) is a manageable number to assay at once.
4. Take the slides through the following steps, all carried out at room temperature:
   a. Fix with 4% paraformaldehyde in 0.1 *M* PBS for 10 min (*see* **Note 13**).
   b. Wash in 1X PBS for 5 min.
   c. Repeat **step 4b** using fresh PBS.
   d. Acetylate with ethanolamine solution for 10 min (to reduce nonspecific binding of the negative probe to the positively charged glass slides and tissue) (*see* **Note 13**).
   e. Dehydrate through ascending 2-min steps of ethanol containing 0.3 *M* ammonium acetate, from 70%, 80%, 90%, then 100% ethanol.
   f. Wash in chloroform for 2 min (to delipidate the sections, thus reducing nonspecific hybridization to white matter) (*see* **Note 13**).
   g. Wash in 100% ethanol followed by 90% ethanol containing 0.3 *M* ammonium acetate.
5. Dry slides with a hairdryer (set to cold air).
6. Place slides (section side up!) In a sealable hybridization container, containing thin foam or filter paper saturated with soaking solution (i.e., 1 part 4X SSC: 1 part de-ionized formamide). There should be sufficient soaking solution to keep the boxes saturated throughout overnight hybridization, but not too much that it spills onto the sections.

Table 3
Prehybridization Steps

Prehybridization treatment	Application
Wax removal	Necessary for *in situ* if the tissue has been preserved by wax-embedding (e.g., in the case of archival tissue).
Fixation	E.g., cross-linked fixatives such as 4% paraformaldehyde) preserve tissue and RNA for *in situ* hybridization. There is an optimal balance between sufficient cross-linking yet still allow probe penetration.
Proteinase K treatment	Proteolytic digestion unmasks nucleic acid targets from cross-links formed in fixation and aids access of long probes in *in situ*. There is a balance between sufficient digestion to allow probe access, but not too much that the protein structure will be weakened and the mRNA lost in solution.
Acetylation	Reduces nonspecific binding in *in situ* of negatively charged probe to positively charged glass and tissue.
Prehybridization	For Northern blotting, before hybridization it is essential to equilibrate with hybridization solution minus the probe to block nonspecific binding sites. This step is not so crucial with *in situ*.

7. Hybridize immediately (we have not found it necessary to prehybridize with hybridization solution minus probe before the hybridization step).

3.1.4. Hybridization

1. Calculate the volume of labeled probe required to give 0.5–2.5 × 10^6 cpm/section (20,000–60,000 cpm/μL) (*see* **Note 14**).
2. Thoroughly mix 2X hybridization solution with an equal volume of de-ionized formamide (*see* **Note 15**) to give the required volume of hybridization buffer (final concentration of 1X hybridization solution, containing 50% (v/v) formamide).
3. Add the calculated amount of labeled probe to the hybridization buffer. Make sure the solution is well mixed and contains no air bubbles. Equilibrate this to 60–70°C for 10 min to denature the probe, then immediately cool it on ice for 2–3 min, to keep the probe single-stranded.
4. Add 10 μL of 1 *M* DTT/mL hybridization mixture, to give a final concentration of 10 m*M* DTT (*see* **Note 6**), mix well and spin down to reduce air bubbles.
5. Pipet the determined aliquots of this hybridization buffer onto each section and, using forceps, gently coverslip with a piece of parafilm cut to the size of the section to prevent dehydration. It is important to cover the entire section with buffer, without scoring the section with the pipet tip or creating air bubbles, which are easily produced by excess pipetting as a result of the BSA in the solution, or by dropping the coverslip over the section.
6. Seal the hybridization chamber with tape, incubate overnight in an oven at the hybridization temperature. We successfully use 42°C for the $GABA_{B1a}$ oligonucleotides and 60°C for the h5-HT_3R riboprobes (*see* **Note 16**).

3.1.5. Posthybridization Washing

The temperature, wash durations, and SSC concentrations used at this stage depend on the properties of the specific oligonucleotides and riboprobes used (*see* **Note 17**). The conditions described as follows work well for the 50 base $GABA_{B1a}$ oligonucleotides and the 750 base h5-HT_3 receptor riboprobes, and can be used as a guide for other similar probes:

1. Dilute 20X SSC to the dilutions required below, and equilibrate these wash solutions in a water bath to the necessary wash temperatures (takes approx 1 h). Sterile conditions do not need to be maintained at this stage (*see* **Note 1**). Make enough solution to completely immerse the slides. We use approx 500 mL per 1 L beaker, containing 25 slides. For the riboprobe washes, also equilibrate RNase solution (25 μg/mL) to 37°C.
2. After hybridization, stack the slides into slide racks and place them in slide boxes, containing 2XSSC at room temperature. Wash the slides for 5 min with slight agitation and, using forceps, carefully remove the Parafilm coverslips.
3. Meanwhile, rinse the used hybridization boxes, then soak them in Decon overnight to reduce radioactive contamination, remembering that everything that

comes into contact with the hybridisation solution is radioactive and should be handled accordingly (*see* **Note 4**).

4. Once all the coverslips have been teased away from the sections, suspend slides in 1-L beakers and wash as follows:
   For the 50 base $GABA_{B1a}$ oligonucleotide: 1X SSC, 0.2% Nathiosulfate at 55°C for 60 min (changing the buffer after 30 min), followed by 0.1X SSC, 0.2% Nathiosulfate at room temperature for 5 min and, finally dH_2O at room temperature for 2 min to remove salts.
   For the h5-HT_3R riboprobe: 2X SSC at 55°C for 30 min, followed by 25 µg/mL RNaseA solution at 37°C for 60 min (*see* **Note 18**), followed by 2X SSC at 50°C for 60 min and finally 0.1X SSC, containing 14 m*M* of 2-mercaptoethanol at 50°C for 3 h, then leave to cool to room temperature overnight (*see* **Note 19**).
5. Dehydrate the sections in 50% ethanol, containing 0.3 *M* ammonium acetate for 4 min, then 70% ethanol, containing 0.3 *M* ammonium acetate for 2 min and finally in 90% ethanol, containing 0.3 *M* ammonium acetate for 2 min (*see* **Note 20**).
6. Dry the slides overnight at room temperature under a paper towel to minimize dust, which may cause background problems if one is emulsion-dipping the sections.
7. Slides are now ready to expose to emulsion to obtain cellular resolution. Alternatively, slides can be exposed to film for rapid signal detection without cellular resolution for quick optimisation of the assay parameters (*see* **Subheading 3.2.6.**).

3.1.6. Probe Detection

3.1.6.1. Emulsion Dipping

1. In the darkroom, equilibrate a water bath to 43°C and for accuracy measure out, in a separate measuring cylinder, the aliquot of dH_2O required to dilute the emulsion one part emulsion to two parts water, knowing that 15 mL of diluted emulsion will coat approx 10 slides (*see* **Note 21**).
   Under safelight conditions (e.g., using Ilford 902–904 safelight with a 15-W bulb):
2. Let the emulsion reach room temperature before removing an aliquot with a clean metal spatula (*see* **Note 22**). Melt this aliquot at 43°C in a measuring cylinder (takes approx 1 h) and in this time also allow the premeasured dH_2O aliquot to reach 43°C.
3. Slowly add the water to the emulsion, pouring the water carefully down the side of the measuring cylinder to prevent air bubbles forming in the emulsion, which will cause uneven coating. Gently pour the diluted emulsion into a slide mailer box (again avoiding air bubble production). Support the mailer in the water bath at 43°C and allow the emulsion to settle for a few minutes.
4. Dip each slide singly into the emulsion while holding the top, labeled end of the slide. Use a uniform dipping technique (e.g., hold each slide in the emulsion for 2 s and slowly extract) to obtain an even emulsion coating of similar thickness over the whole of each slide.

5. Blot the bottom and underside of the slide on a paper towel, and lie the slides flat on a metal tray cooled on ice, or on a cold plate, for 2 h in total darkness to set the emulsion.
6. Once the emulsion has set, the slides can be slowly dried vertically overnight in total darkness.
7. The next morning, pack the dry slides in slide boxes, containing silica gel desiccant, seal with electrical tape, wrap in foil and then black plastic bag, and store the boxes at 4°C in the dark, away from any source of radiation or strong chemicals for required exposure time (this being 8 to 10 wk for the PPD oligonucleotide and h5-HT_3R riboprobe experiments). Initially, it is a good idea to prepare a number of similarly treated ISHH slides and develop these at different time points to ascertain the optimal exposure time for that specific probe (*see* **Note 23**).

3.1.6.2. Emulsion Developing

1. Remove the boxes of emulsion-coated slides from the cold room and equilibrate to room temperature (takes approx 30 min) before opening them, to prevent condensation forming on the slides, which may wrinkle the emulsion coat. The slides should be treated gently at all times, as the emulsion coat is very prone to mechanical stress and is easily scratched under safelight conditions (using ilford 902–904 safelight with a 15-W bulb).
2. Carefully remove the slides from their box, put them in slide racks, and process as follows, with gentle agitation (checking before hand that the temperature of the following solutions is below 20°C, as silver grain size is proportional to temperature):

Developer (at 18°C)	4 min
dH_2O	rinse
Fixer	4 min (this being twice the time it takes emulsion to clear)
Fixer	4 min
dH_2O	5 × 10 min (*see* **Note 24**)

3. Stain and mount slides immediately.

3.1.6.3. Histological Staining of Slides Using Hematoxylin/Eosin

Aim to obtain a light stain, so the blue color does not interfere with image analysis (*see* **Note 25**). Stain the slides as follows:

1.	Mayer's Hematoxylin	20 s
2.	dH_2O	5 s
3.	Alkaline H_2O	30 s
4.	dH_2O	30 s
5.	70% EtOH	1 min (needed because alcohol-based eosin is used)
6.	1% eosin Y	approx 1 s (dilute if too "young")

7.	Acid alcohol (70%)	≥15 s (this takes out excess eosin)
8.	95% EtOH	Twice 1 min
9.	100% EtOH	Twice 1 min
10.	Histoclear	Twice 2 min (clearing problems occur with "old" Histoclear)

11. Immediately mount the sections in DePeX mounting medium and gently coverslip with the aid of forceps to avoid air bubbles (*see* **Note 26**).

3.1.7. Controls

3.1.7.1. Negative Controls

1. Antisense vs sense probes: replace the antisense probe with a labeled sense probe, which has a complementary sequence to that of the antisense (i.e., an identical sequence to the mrna under investigation) and therefore will have similar physical properties to the antisense probe, but should not hybridize under identical assay conditions (*see* **Notes 3** and **7**, and **Fig. 1A–C** vs **E**).
2. RNaseA pretreatment of tissue: after prehybridization **Subheading 3.1.1.3.**), pipet 100 µL of RNase buffer containing RNaseA (1 µg/µL) onto each section and incubate the slides in sealed hybridization boxes for 60 min at 37°C. After this time, tip off the excess solution, wash the slides twice in excess DEPC-treated dH_2O at room temperature for 5 min each time, dry the sections gently with a hairdryer (set to cold), and carry out hybridization as described in **Subheading 3.1.4.** Under these conditions, no signal should be detected (*see* **Fig. 1A–C** vs **D**). This control is more relevant for oligonucleotide and cDNA probe-ISHH, where it is not so critical to the probes if some RNaseA activity remains during hybridization. *See* **Note 27**.

3.1.7.2. Positive Controls

1. If possible, it is useful to test the ISHH protocol on cell lines highly expressing the signal of interest.
2. Confirm the identity of the detected RNA species by molecular weight determination, using Northern hybridization (*see* **Fig. 2**).
3. Assay tissue regions known to discretely express the signal.
4. Check that the obtained pattern of expression is anatomically "sensible" with published data. *See* **Note 27.**

3.1.8. Cellular Analysis

Positively labeled cells are examined and photographed by bright- and dark-field microscopy and results interpreted with respect to results obtained from control experiments (*see* **Subheading 3.1.1.7.** and **Note 28**).

3.1.8.1. CELL COUNTING

1. Count cells as being positively labeled under the light microscope at 400× magnification, using the following criteria for [^{35}S]-labeled probes:
   a. Silver grains show a halo pattern around a distinct nucleus surrounded by a pale pink stained cytoplasm
   b. The halo pattern is at least five times more dense than the background
   c. Grains representing mRNA-specific hybridization will be on a different focal plane to nontissue-derived or tissue surface-derived general background.

   See **Fig. 1**.

3.1.8.2. SILVER GRAIN COUNTING

1. The silver grain density is measured with the aid of an automated image analysis system (such as Improvision "IMAGE" software on an Apple Macintosh II computer).
2. Measure the background silver grain density in an adjacent area to positive cells and subtract this from total counts to give a net density per measured area.

3.2. Northern Hybridization

It is important to take precautions against RNase contamination when carrying out the following protocols (*see* **Note 1**).

3.2.1. Probe Labeling

Label probes for Northern blotting in exactly the same manner as described in **Subheading 3.1.1.2.** or **3.1.1.3**. Generally, cDNA and riboprobes are the probes of choice in Northern hybridization, as these are generally more sensitive than shorter oligonucleotides (*see* **Table 1**). Usually [^{32}P] is the label of choice in Northern hybridization, where signal scatter and low resolution are not important (*see* **Table 2**). The following protocols describe the successful use of [^{32}P]-labeled h5-HT_3R riboprobes. Standardly, the probe is labelled one morning and used that evening in overnight hybridization.

3.2.2. RNA Collection

3.2.2.1. TISSUE COLLECTION FOR RNA ISOLATION

1. Immediately after death, isolate the tissue of interest, using sterile instruments and working as quickly as possible to reduce RNase activity and RNA degradation.
2. Wrap the specimen in silver foil and snap-freeze it for 5 min in liquid nitrogen.
3. Store the frozen tissue in sealed water-tight containers at –70°C.

3.2.2.2. Rapid Extraction of Total RNA

This method is based on that by Chomczynski and Sacchi *(23)*.

1. In a fume hood, wash a homogeniser or sonicator probe with 3% H_2O_2 solution, followed by 70% ethanol solution, then 0.1 *N* NaOH solution containing 1 mm EDTA, and finally with at least seven changes of DEPC-treated dH_2O, to remove any possible rnase contamination.

The following procedures must be carried out in a cold room or on ice, using prechilled solutions:

2. Weigh out 30–40 mg tissue into autoclaved microcentrifuge tubes, using a sterile scalpel blade to cut samples from the frozen tissue block (*see* **Note 29**).
3. Immediately add 400 µL of GTC denaturing solution.
4. Homogenize the tissue on ice for approx 4 s, washing the homogenizer as **step 1** before each new sample.
5. Add 40 µL 2 *M* sodium acetate (pH 4.0) and vortex well.
6. Add 400 µL TE saturated phenol and vortex (the bubbles should disappear here).
7. Add 80 µL chloroform:isoamyl alcohol mix (49:1). (The presence of isoamyl alcohol gives a sharper and more hydrophobic interface, allowing better visualization and more efficient removal of the aqueous phase.)
8. Vortex for at least 10 s to obtain an emulsion, then cool on ice for 15 min.
9. Spin for 20 min at 4°C at 13,400*g* in a microcentrifuge.
10. Transfer the aqueous phase (i.e., the top, very clear layer of approx 400 µL in volume) to a clean microcentrifuge tube, taking care not to contaminate this with any of the protein interface, which will result in impure RNA or with the lower phenol-chloroform phase, which may prevent rna precipitation.
11. Add 1 volume (i.e., approx 400 µL) of isopropanol and precipitate the RNA at –20°C for at least an hour (overnight for maximum recovery).
12. Spin for 20 min at 4°C at 13,400*g* in a microcentrifuge.
13. Carefully remove and discard the supernatant with an autoclaved glass Pasteur pipet. A small opaque pellet of RNA should now be visible at the bottom of the microcentrifuge tube.
14. Flick resuspend the pellet in 120 µL of GTC denaturing solution.
15. Add 1 vol (i.e., 120 µL) of isopropanol and re-precipitate the RNA at –20°C for at least an hour (overnight for maximum recovery).
16. Spin for 10 min at 4°C at 13,400*g* in a microcentrifuge.
17. Remove the supernatant with an autoclaved drawn out Pasteur pipet, taking care not to touch and therefore shear the RNA pellet.
18. Wash the pellet by flicking it in at least 400 µL of 75% ethanol made with DEPC-treated dH_2O.
19. Spin for 10 min at 4°C at 13,400*g* in a microcentrifuge, then pour off the ethanol.
20. Dry the pellet inverted on the bench at room temperature for about 10 min; long enough to remove the ethanol but not too long that the over-dried pellet will be difficult to resuspend.
21. Resuspend in 40 µL DEPC-treated dH_2O (i.e., 1 µL/mg starting tissue).

22. Store at –20°C ready for Northern blotting (or for poly(a)$^+$ RNA extraction; *see* **Note 30**).
23. Measure the absorbance values of the total RNA solution, using a quartz cuvet and a UV spectrometer at wavelengths of 230 nm (absorbance of guanidine thiocyanate), at 260 nm (absorbance of nucleic acids), and at 280 nm (absorbance of protein) to calculate:

    a. purity, where
       $A_{260/280}$ ratio = 2.0 represents 100% purity of RNA to protein
       $A_{260/230}$ ratio >2.0 indicates successful removal of GTC and 2-mercaptoethanol
    b. concentration, where (for a pathlength of 1cm)
       1 unit at Abs_{260} = 40 µg/mL RNA (expect a yield of 1–2 µg total RNA/mg starting tissue)
24. Measure the 28S:18S ratio on a denaturing gel to obtain a measure of the RNA integrity (*see* **Note 31**).

3.2.2.3. RNA Gel Fractionation

1. Soak a suitably sized gel tray, stops, and well molds that will hold 20 µL volumes in DEPC-treated dH_2O for at least one hour (*see* **Note 32**).
2. Make enough agarose (1%)/formaldehyde solution to give a gel no more than 0.5 cm thick.
3. Set up the gel case in a fume hood, pour the cooled gel into the RNase-free gel tray, allow this to set (takes 15–30 min), then remove the stops and "age" the gel in 1X MOPS running buffer for at least 15 min.
4. Meanwhile, thaw the RNA samples and an aliquot of de-ionized formamide on ice.
5. Prepare fresh loading solution by mixing 100 µL of 10X MOPS running buffer with 175 µL of formaldehyde and 500 µL de-ionized formamide.
6. For each RNA sample, dilute and gently mix (by flicking) up to 20 µg of RNA (at least 5 µg) in loading solution to give a final volume of 20 µL. Similarly prepare 20 µL of RNA marker solution, containing 3 µg RNA markers diluted in loading solution.
7. Incubate the diluted samples at 65°C for 15 min to denature the RNA, then rapidly cool them on ice to keep the RNA denatured.
8. Add 1 µL of loading dye to each 20 µL of diluted sample and carefully pipet the aliquots into separate wells of the "aged" gel, noting which order the samples and markers are loaded. Take care not to pierce the bottom of the gel or to expel the last drop of solution from the pipet tip, as this may cause air bubbles to push the solution out of the well. Work quickly to minimize sample diffusion.
9. Immediately run the gel at 100 V (approx 100 ma) from the anode to the cathode for 3 h or until the loading dye front is at least two-thirds down the gel.
10. Chop off the markers with a sterile scalpel blade and incubate these with approx 2 µg ethidium bromide/mL running buffer for 20 min, destain overnight in DEPC-treated dH_2O at 4°C to remove formaldehyde and excess ethidium bromide in

order to visualize the RNA on a UV transilluminator, and process as described in **Suhbeading 3.2.8**.

11. Trim off any excess gel from around the edges (to economize on blotting materials in later steps) and remove the top right corner so the gel orientation can be identified.
12. Soak the gel containing the samples for 20 min in 0.05 *N* sodium hydroxide, rinse in DEPC-treated dH_2O, soak for 45 min in 20X SSC and immediately transfer the RNA to membrane by Northern blotting.

3.2.2.4. Northern Blotting by Capillary Transfer

See **Fig. 3**.

1. Cut a piece of nylon membrane (e.g., hybondn+, Amersham Biosciences, http://www.amershambioscinces.com) to the exact size as the gel, chop the top right corner to match the gel and label the membrane. Handle the membrane very carefully to avoid putting any pressure on it or touching it without gloves, both of which will result in increased background levels.
2. Cut three pieces of 3-MM Whatman chromatography paper to use as a wick.
3. Cut three pieces of 3-MM Whatman chromatography paper to the exact size as the gel.
4. Pour 20X SSC (approx 1 L) into a plastic trough.
5. Wet the three paper wicks and place these over a plastic tray, bridging the trough so that each end of the wick is in the 20X SSC solution.
6. Roll out the wick, using a Pasteur pipet like a rolling pin, to remove any air bubbles which will cause uneven flow of solution.
7. Place the gel in the center of the wick covered bridge.
8. Wet the nylon membrane for 1 min in 20X SSC, then place this to fit exactly on top of the gel, lining it up with the top of the wells. Roll out any air bubbles, using a Pasteur pipet as before, and finally mark the positions of the wells on the membrane (*see* **Note 33**).
9. Dip the three pieces of gel-sized 3-MM Whatman paper for 1 min in 20X SSC then place these to fit exactly over the membrane. Roll out any air bubbles, using a Pasteur pipet as described in **step 6**.
10. Surround the membrane with Parafilm to seal it fully, preventing solution flow from anywhere in the trough other than through the gel.
11. Cross four pieces of folded absorbent paper over the membrane set up at angles of 0°, 90°, 180°, then 270°.
12. Place an approx 5-cm thickness of paper towels over the absorbent roll, put a flat tray over these and a 500 g weight on top of that (e.g., a 500-mL bottle of water) to aid consequential RNA transfer from the gel to the membrane with solution flow.
13. Leave this set up for at least 18 h at 4°C, replacing wet paper towels with fresh ones as necessary.

14. The next day, dismantle the set-up and carefully lift the blotted membrane from the gel. Wash the membrane briefly in 2X SSC to remove any agarose and air dry for 30 min between filter paper (*see* **Note 34**).
15. Bake the membrane blot between two sheets of 3-MM Whatman chromatography paper for 2 h at 80°C to permanently immobilize the RNA on to the membrane (*see* **Note 35**).
16. Hybridize the membrane blot immediately or mount it on fresh filter paper, seal in cling film, then in foil and keep at 4°C until hybridization. By doing the latter, a bank of membranes can be set up and stored for several weeks before hybridization.

3.2.3. Northern Prehybridization

Prehybridization is an important step for reducing background levels in Northern hybridization.

1. Make the prehybridization solution as below, in a sterile 15-mL vial and equilibrate this to the hybridization temperature (this being 55°C for the h5-HT_3R riboprobes [*see* **Note 36**]).

Prehybridization solution:	*For 10 mL*	*Final*
De-ionized formamide	5 mL	50%
20X SSPE	2.5 mL	5X
50X Denhardt's solution	1 mL	5X
10% SDS	500 µL	0.5%
0.5 *M* EDTA	20 µL	1 m*M*
DEPC-treated dH_2O	880 µL	–
Salmon sperm ssDNA	100 µL	100 µg/mL

2. Wet the mesh support in a trough of 2X SSPE at room temperature and lay the membrane blot over this, remembering to handle this carefully and only by the corners. Wet the membrane, roll it in the mesh, and place the whole roll in the 50 mL hybridization vial with enough (approx 15 mL) 2X SSPE to help roll out mesh around the inside surface of the vial, avoiding all air bubbles (which, if not removed, will prevent even distribution of solution and cause hot spots of background radioactivity over the membrane).
3. When the membrane is in place, wrapped around the perimeter of the inside of the vial with no air pockets, replace the 2X SSPE with 5 mL of prewarmed prehybridization solution.
4. Prehybridize at 55°C for 1.5 h, with rotation.

3.2.4. Northern Hybridization Using [^{32}P]-Labeled Riboprobes

1. During prehybridization, prepare the hybridization solution as follows, in a sterile 15 mL vial.

Hybridization solution	*For 10 mL*	*Final*
De-ionized formamide	5 mL	50%
20X SSPE	2.5 mL	5X
10% SDS	100 µL	0.1%
0.5 *M* EDTA	20 µL	1 m*M*
DEPC-treated dH_2O	2.28 mL	–
Salmon sperm ssDNA	100 µL	100 µg/mL

2. We use all the labeled product from one labeling reaction of 50 µCi [^{32}P] (i.e., at least 16 ng of riboprobe of up to 1–2 × 10^8 dpm; **Subheading 3.1.1.3.**) for 5 mL of hybridization solution and one membrane blot, and perform one antisense and one sense reaction simultaneously in two separate vials. Denature the aliquot of [^{32}P]-labeled riboprobe at 60–70°C for 10 min then rapidly cool it on ice. Add this denatured probe to the hybridization solution, mix thoroughly, and equilibrate this mixture to the hybridization temperature in a water bath (*see* **Notes 4** and **36**).
3. Immediately after prehybridization, replace the solution in the vial with the 5-mL aliquot of prewarmed hybridization solution.
4. Hybridize at 55°C, overnight with rotation.

3.2.5. Posthybridization Washing

1. Pre-equilibrate the washing solutions to their correct temperatures (*see* **Note 17**).
2. Pour off the radioactive hybridization solution (*see* **Note 4**).
3. Wash the membrane (50 mL/wash) with rotation, in:
   a. 2X SSC, containing 0.1% SDS for 15 min at room temperature
   b. 2X SSC, containing 0.1% SDS for 15 min at 60°C
   c. 2X SSC, containing 0.1% SDS for 15 min at 60°C
   d. 0.1X SSC, containing 0.1% SDS for 15 min at 60°C
   e. 0.1X SSC, containing 0.1% SDS for 15 min at 60°C

   Discard the wash solution and assess the amount of radioactivity left on the membrane with a hand-held β-counter to adapt the stringency of the next wash accordingly, before adding the next solution (*see* **Note 37**).
4. Air-dry the membrane on filter paper at room temperature behind protective shielding for 20–30 min.
5. Mount the membrane blot by the corners onto fresh filter paper, cutting the same corner of the filter paper as the membrane and label before covering with plastic wrap and then exposing to autoradiography film.

3.2.6. Probe Detection

3.2.6.1. Exposing to Autoradiography Film

In the darkroom, using Ilford 902-904 safelight and 15-W bulb:

1. Cut one corner of the autoradiography film (usually the same corner as the membrane to avoid confusion) to identify the orientation of the film after development. Arrange the film in a film cassette and fix securely in position with tape.

2. Tape the mounted membrane blot securely to the film, avoiding creases in the plastic wrap.
3. Mark the exact position of the wells on the film to enable accurate calculation of the size of the detected bands (*see* **Subheading 3.2.8.**).
4. Seal the cassette against light, label, date and store it at –70°C. For the h5-HT_3R riboprobes, the film required exposure for 4–5 d. The blot can be re-exposed to film if this time period is not sufficient, or intensifying screens may be used to decrease the exposure time.

3.2.6.2. Film Developing

1. Remove the film cassette from the –70°C freezer and allow it to equilibrate to room temperature for up to 1 h before opening the cassette under safelight conditions.
2. Develop the film as in **step 2** of **Subheading 3.1.6.2.**
3. Hang the film up until dry, then visualize the silver grains on a light box (*see* **Note 38**).

3.2.7. Controls

3.2.7.1. Negative Controls

1. Antisense vs sense probes: replace the antisense probe with a labeled sense probe, which has a complementary sequence to that of the antisense (i.e., identical sequence to the mRNA under investigation), and therefore will have similar physical properties to the antisense probe, but should not produce any bands under identical hybridization conditions (*see* **Notes 3** and **7**).
2. Assay RNA extracted from a source known to be devoid of the RNA species under investigation.

3.2.7.2. Positive Controls

1. The size, number of bands, and selectivity of the signal, as compared with the pattern obtained with the sense strand, give good indications of the validity of signal detection (*see* **Fig. 2**).
2. Assay RNA extracted from a cell line highly expressing the mRNA of interest and/or from an area known to express the RNA species in abundance (*see* **Fig. 2**).
3. Re-hybridize the stripped blot with a probe to a constitutive RNA species (such as β-actin mRNA), which is expressed in high and constant amounts independent of external influences, as an internal control to check the integrity and the amount of each RNA sample loaded on the gel. This also provides a method to quantify any changes in expression in the mRNA of interest under different conditions.

3.2.8. Analysis

1. Visualize the ethidium bromide-stained markers on the destained gel (**Subheading 3.2.2.3., step 10**) on a UV transilluminator (taking the necessary precautions to protect your eyes from UV light) and photograph these against a ruler.

2. Mark off the position of each molecular weight marker band on graph paper and plot the log of the molecular weight (given on the product data sheet) against band migration distance from the well (in mm) to obtain a linear standard molecular weight calibration graph.
3. After the film has been developed (**Subheading 3.2.6.2.**), measure the migration distance (in mm) of detected bands on the film from the top of the marked well position. Calculate the molecular weight of these bands, using the equation of the line for the linear standard molecular weight calibration graph obtained in **step 2**.

4. Notes

1. It is critical to maintain RNase-free conditions prior to and during hybridization. To minimize RNase contamination, bake glassware overnight at 200°C and autoclave pipet tips, microcentrifuge tubes, solutions and so on where possible. If it is not possible to autoclave or bake items, these should be sterilized with 70% ethanol, then rinsed thoroughly with DEPC-treated dH_2O. Always wear gloves and avoid breathing directly on RNase-free items, as RNase is present on skin and in breath. Make solutions with DEPC-treated dH_2O where possible or autoclaved dH_2O if not (DEPC cannot be added directly to Tris-containing solutions, as primary amines will be produced). RNA/RNA and DNA/RNA hybrids are RNase-resistant, so non-RNase-free procedures can be carried out after hybridization.
2. RNase-free spun columns are commercially available (e.g., IBL Nuclean D25 columns; store at 4°C, or Biospin 30 [Bio-rad 732-6004] for DNA >20 bp). However, it is cheaper and quite straightforward to make them in-house, provided that RNase-free conditions can be maintained.
3. To design a suitable antisense oligonucleotide probe, select a complementary area within the transcribed sequence of interest of 20–50 bases, which is selective for that sequence when compared with all other known gene sequences on a database (using, for example, a FASTA or BLAST search) and has a GC content of 50–60% (*see* **Subheading 1, item 4**). It is optimal to have 100% bp homology with the mRNA sequence, as only one mismatch in a short probe may be enough to lose signal. cDNA sequences for $GABA_{B1a}$, $GABA_{B1b}$, and $GABA_{B2}$ were acquired from the Entrez Nucleotide QUERY database given the reference numbers of the published sequences (24). These cDNA sequences were screened for homology against the rest of the database using a FASTA search, and the results were used to select candidate sequences of 50 $GABA_{B1a}$ nucleotides that were subsequently screened for homology to other cDNAs on the database also using a FASTA search.

   The sense strand has the identical sequence to the mRNA in this region, and therefore has similar physical properties to the antisense strand (i.e., a length of probe, GC content, Tm value, and molecular weight), yet should not hybridize. Shown as follows is the antisense probe sequence used to detect the $GABA_{B1a}$ mRNA, running left to right from the 5'-end to the 3'-end:

$GABA_{B1a}$
CAA-ATA-AGA-CTT-GGA-GCA-GAT-TCG-GAC-ACA-GCG-GCT-GGG-TGT-GTC-CAT-AT

$$\text{Mol wt} = (251 \times nA) + (227 \times nT) + (267 \times nG) + (242 \times nC) + ([62 \times n-1]) + (54 \times n) + (17 \times [n-1])$$

where: nA, nT, nG, nC = number of respective bases in the probe sequence
$62 \times [n-1]$ = molecular weight of phosphate groups
$54 \times n$ = molecular weight of water molecules/nucleotide
17 x n–1 = mol wt of ammonium cations associated with the phosphate groups.

The stock concentration of antisense used in our case was 1 μg/μL, from which aliquots were diluted at 150 ng/μL with autoclaved DEPC-treated dH_2O ready to be used directly in the reaction mix.

4. Precautions should stringently be adhered to when working with radioactive isotopes such as [^{35}S] and [^{32}P]. These precautions include the use of protective acrylic plastic shielding in the case of [^{32}P] isotopes, regular radiation level monitoring of persons and the designated working area, avoidance of aerosol production, and disposal of radioactivity as required by the institute's regulations. Use only fresh radiolabel with a high specific activity. This will produce a probe of high specific activity and lower background.
5. Some oligonucleotides label better than others, and this appears to depend on their base composition. For good 3' end labeling, we find that a stoichiometry within the reaction mix of approx 30 pmol of [^{35}S]-αdATP to 1 pmol oligonucleotide probe and a TdT enzyme dilution of at least 1 in 10, works well. [^{32}P]-labeled oligonucleotide probes can be generated by replacing [^{35}S]-αdATP with a similar molar quantity of [^{32}P]-αdATP in the labeling reaction and carrying out the procedure in the same way.
6. The reducing agent DTT is added to stabilize nucleic acid hybrids and also to prevent the formation of disulphide bridges in [^{35}S]-labeled probes, keeping the probe single-stranded and thus available for hybridization.
7. A suitable antisense riboprobe can be 50–1000 bases long, depending on the tissue type and the way in which this tissue has been fixed and pretreated, all of which affect the degree of probe access to hybridization sites. When designing a suitable vector to generate a riboprobe, select from the restriction digest map, an area within the transcribed sequence of interest which can be subcloned into an appropriate vector expression system for in vitro transcription, and which has a high GC content (approx 50%, but not too high that the probe will be very "sticky," causing background problems) and is selective for that sequence when compared with a database containing all other known gene sequences, using for example a FASTA or BLAST search. Once this selected cDNA sequence has been subcloned, the resulting vector is amplified and purified. Linearized cDNA template is then produced for in vitro transcription by cutting the vector containing the subcloned region

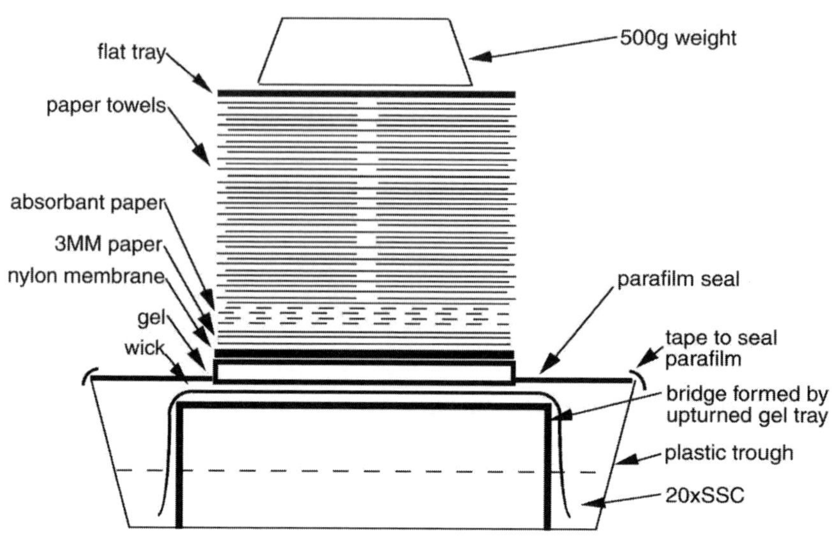

Fig. 4. Apparatus for Northern blotting by capillary transfer.

of cDNA sequence with an appropriate restriction enzyme and purifying the product. An appropriate restriction enzyme is one which will cut the sequence either immediately downstream of the cDNA insert or within the insert, yet leave the promoter site intact so that specific sized transcripts, which contain minimal nonspecific vector sequence, are generated. **Figure 4** shows a schematic representation of a vector containing cDNA corresponding to amino acid residues 62 to 312 of the h5-HT_3 receptor sequence. The pBluescript II SK^+ plasmid vector is selectively amplified in *Escherichia coli* XL1-Blue MRF' competent cells (Stratagene) in the presence of ampicillin. To generate antisense transcripts, the resulting purified plasmid is cut with KpnI and transcription performed from the T3 RNA polymerase site. To generate sense transcripts, the vector is linearized with BamHI and transcription performed from the T7 promoter site. For convenience, a stock of linearized cDNA template is stored at –70°C and when necessary, an aliquot is diluted in DEPC-treated dH_2O to the required concentration for *in vitro* transcription. Aim to add 0.5–1.0 µg of template to the reaction set out in **Subheading 3.1.3.** It is best to avoid using restriction enzymes which produce 3'-overhang ends on the cDNA template. These ends can act as promoters to initiate nonspecific or wrap-around transcripts. If 3' sticky ends are unavoidable, such as when using KpnI to generate antisense h5-HT_3R riboprobes, it is necessary to blunt end the cDNA with T4 DNA polymerase before in vitro transcription.

Add 5 U of T4 DNA polymerase (i.e., 0.5 µL of 10 U/µL stock) per µg cDNA template to the reaction mix containing transcription buffer, DTT, rRNasin, and linearized DNA, as detailed in **Subheading 3.1.1.3., step 1**. Incubate at 22°C for 15 min, then add the remaining ingredients and proceed with in vitro transcription exactly as described.

As an alternative to subcloning, the region of the cDNA of interest can be amplified by polymerase chain reaction (PCR) with two synthetic oligonucleotides, each containing a 5' extension corresponding to promoter sequence of SP6 (sense primer) or T7 (antisense primer) RNA polymerase. The PCR fragments can be then purified, precipitated, resuspended and used as templates for synthesizing radiolabelled sense or antisense riboprobes (*see* **ref. *25***).

8. Addition of cold UTP to the reaction mix should increase the amount of transcription, but will lower the specific activity of the resultant probe. We find that this also dramatically increases nonspecific binding and therefore we avoid adding cold UTP.
9. [^{32}P]-labeled riboprobes can be generated by substituting [^{35}S]-αUTP for 50 μCi of [^{32}P]-αUTP and proceeding as detailed.
10. RNA polymerases are very labile and should be kept on ice and returned to the freezer immediately after use. No more than 10 U/μg cDNA template is required, as promoter-specificity will be lost if excess polymerase is used; at high concentrations, T7 RNA polymerase may act at the T3 promoter site and vice versa.
11. It is possible to buy RNase-free subbed or positively charged slides (e.g., BDH Superfrost plus microscope slides). These are very convenient to use and competitively priced.
12. Throughout the cutting session, always wear gloves and avoid breathing over tissue to avoid RNase contamination (*see* **Note 1**). Similarly, a new disposable blade should be used for each new sample. This blade is initially cleaned with xylene followed by ethanol, then frequently re-cleaned with ethanol throughout the cutting session.
13. The exact steps employed in prehybridization treatment may vary depending on the nature of the tissue: the length of the paraformaldehyde fixation step is critical and times may need to be optimized to produce sufficient tissue fixation without causing excessive cross-linkage which will inhibit probe penetration. We have standardly acetylated, then delipidated central nervous system (CNS) sections with chloroform, but these steps may be ineffective in other situations and on other tissues.
14. We successfully use 20 μL of 1X hybridization buffer/rat spinal cord section, 50 μL of 1X hybridization buffer/rat coronal brain section, and 200 μL of hybridization buffer/human brain section (approximately 4 cm^2). The volume of hybridization buffer used will obviously depend on the size of the section. Aim to have minimum volume and therefore maximum probe concentration, yet sufficient solution that the section is completely covered and will not dry out. The optimal probe concentration may differ between probes, therefore initially, it may be necessary to try a range of probe concentrations to find which is best for the particular application.

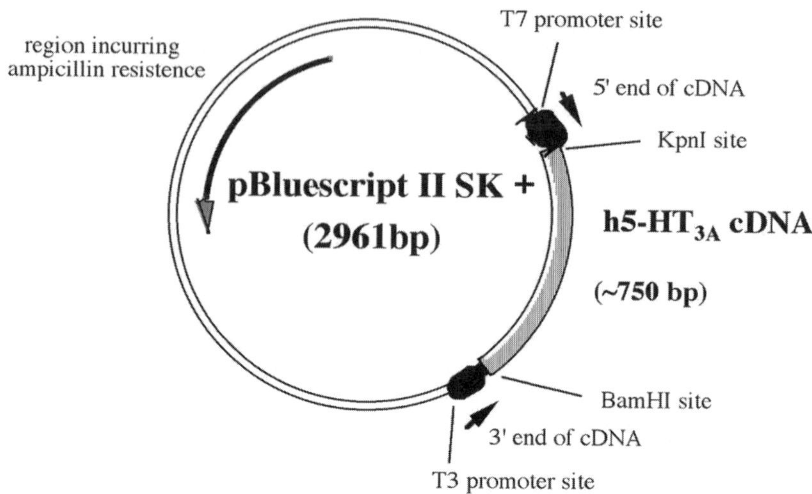

Fig. 5. Construct used to generate h5-HT_3 receptor antisense riboprobe.

15. Formamide is a hydrogen bond breaker, therefore it acts as a destabilizing agent, reducing nonspecific binding. The signal-to-noise ratio can be manipulated by adjusting the percentage of formamide in the hybridization buffer (*see* **Subheading 1, item 4**).
16. The hybridization temperature is dependent on the properties of the probe used, as defined by the Tm equations (**Subheading 1, item 4**). These equations reveal how hybrids formed between RNA and riboprobes are more stable than those formed with short DNA oligonucleotides, therefore formation of the former can withstand more stringent hybridization conditions.
17. The rate of hybridization increases with increasing salt concentration and decreasing temperature, therefore the ratio of hybridization to nonspecific background can be increased by adjusting the temperature and/or the salt concentration at the posthybridization wash stage. The wash conditions for oligonucleotides are usually much less stringent than when using a riboprobes, as DNA:RNA hybrids are less stable than RNA:RNA hybrids (*see* **Subheading 1, item 4**). However, if these conditions are too stringent the probe will be stripped off completely, yielding no signal detection.
18. RNaseA destroys ssRNA, leaving duplexed RNA intact. This is therefore an important step in removing any nonhybridized riboprobe and thus reducing background. Keep all containers and solutions containing RNaseA away from everything which may come into contact with materials used prior to and during subsequent hybridization assay.
19. 2-mercaptoethanol (or DTT) is used to stabilize hybridized riboprobe, but also inhibits RNase activity and therefore cannot be added to washes prior to the RNaseA step.

20. The ammonium acetate is added to reduce salt crystal formation, which will cause background problems if the slides are emulsion-coated.
21. Glassware used in emulsion dipping (e.g., measuring cylinders) should be washed in chromic acid before use to remove any emulsion remaining from the previous dipping session, which may lead to increased background levels.
22. The emulsion has a short shelf life and should only be opened approximately five times before excessive background levels may become a problem. Mechanical stress will also increase background, so the melted emulsion and coated slides should be handled gently throughout the procedure.
23. As a control for the emulsion dipping procedure, it is useful to process a blank subbed-slide through the dipping steps, along with the slides being assayed and develop it the next day, when boxing the others. This "test" slide will reveal the evenness of the dipping technique and show if the slides have been exposed to any light/radiation source or excessive mechanical stress at any point through the procedure or dried too rapidly, all of which will increase background levels. Acceptable background, according to the manufacturer's instructions, is 10–25 grains/100X field. Histological staining (e.g., as in **Subheading 3.6.3.**) will also highlight any streaking effects resulting from uneven emulsion coating.
24. It is important that the slides are washed for at least 30 min to remove excess chemicals before they are histological stained. This can be done in tap water, but the final rinse should be in dH_2O.
25. Hematoxylin is a basic blue dye for nucleic acids, whereas eosin stains cytoplasm a pale pink/orange. Other examples of basic blue dyes are cresyl fast violet, toluidine blue, and thionin. In contrast, pyronin can also be used, this giving a pink stain. Preference will depend on which stain gives least interference to the computer aided image analysis. It may be necessary to adjust the staining times set out here depending on age of the stains and the condition of the tissue sections (eosin will be taken up very quickly if the sections are of good quality, so be careful).
26. Glass coverslips can be removed from the sections at a later date by soaking the slides overnight in xylene to dissolve the DePeX mounting medium. The slides should then be rehydrated through decreasing concentrations of ethanol, from 100% ethanol to dH_2O and subsequently re-stained if necessary.
27. Other negative controls which may be employed to knock out specific hybridization, include (a) competition studies with 100-fold excess of unlabeled "cold" antisense probe co-applied with labeled antisense probe, and (b) using nonsense probes, such as scrambled oligonucleotides, which have the same base composition and therefore similar physical properties to the antisense probe, but no complementary sequences for possible hybridization. Other positive controls include (a) reconfirmation of the results when the experiments are repeated with another probe designed to a different region of the same mRNA (*see* **Notes 3** and **7** for design of probes), and (b) probing for a constitutive mRNA, e.g., β-actin, to verify the tissue RNA integrity.

28. Silver grain density is not linearly correlated with amount of RNA, therefore analysis can only really be considered semiquantitative at best. Furthermore, absence of detection may only reflect lack of sensitivity of the protocol and not absence of RNA expression. It may be necessary to use *in situ* PCR to detect very low-expressing mRNA. Conversely, the presence of mRNA does not automatically demonstrate the presence of the translated peptide product.
29. RNA may be extracted from up to 1 g of starting tissue by scaling up the quantities in this protocol proportionally.
30. Protocols for poly(A)$^+$ RNA extraction are not covered in this chapter; however, it is possible to purchase kits to perform this procedure, for example Promega PolyATtract mRNA Isolation Systems.
31. To measure the RNA integrity, run the samples on a denaturing gel, as described in **Subheading 3.2.2.3., steps 1–10**, except do not remove the marker lane. Instead, visualize all the RNA by staining the whole gel with EtBr (**Subheading 3.2.2.3., step 10**). The total RNA should appear as a faint smear through the lane, with the most abundant RNA species (the 28S and 18S ribosomal RNA, which constitutes 80–85% total RNA) appearing as two strong and distinct bands. The 28S, being twice as abundant, should appear twice as strong as the 18S band (mRNA is estimated to make up 1–5% of total RNA). If the RNA has been degraded during isolation, it will run to the bottom of the gel with little or no indication of ribosomal banding.
32. If possible, it is better to dedicate a gel tray and wells solely for RNA work to minimize the risk of RNase contamination and therefore degradation of the RNA samples.
33. The membrane must exactly cover the gel to prevent gel dehydration and maximize solution flow through the gel, and thus subsequently maximize RNA transfer, and also so that the migration distance (and therefore the molecular weight of the detected bands) can be measured as accurately as possible.
34. The percentage transfer of RNA from gel to membrane can be checked by staining the gel with EtBr and visualizing any RNA remaining after blotting on a UV transilluminator. It is better to blot in the absence of EtBr and stain the gel afterwards in order to ensure that transfer has been successful, as EtBr itself may affect RNA mobility.
35. Alternatively, the RNA can be efficiently cross-linked to the membrane with UV light exposure for up to 5 min. However, the exposure time in this method is critical and variable depending on the transilluminator used, so the optimal exposure time has to be calibrated accordingly.
36. High background will result if solutions are not equilibrated to the hybridization temperature, which is dependent on the properties of the probe used, as defined in the Tm equation (*see* **Subheading 1, item 4**).
37. The stringency of the next wash can be adapted appropriately, depending on the level of radioactivity measured on the hand-held counter. In this way, a more informed wash strategy can be applied in each experiment. A high level of radioactivity uniformly spread over the membrane is indicative that a high level of overall background still remains, requiring that further washing steps carried out.

In contrast, the wash procedure should be terminated when membranes show only discrete patches of radioactivity, indicating only specific hybridization remains (or even no signal with the relatively insensitive hand-held counter).

38. The membrane can be stripped and re-probed after the film has been developed. To strip the blot, pour boiling, 0.5% (w/v) SDS solution over the membrane and allow it to cool to room temperature. It is difficult to completely strip the probe from the membrane, but stripping should be sufficient so that the mRNA on the membrane can be re-probed.

References

1. Gall, J. G. and Pardue, M. L. (1969) Formation and detection of RNA-DNA hybrid molecules in cytological preparations. *Proc. Natl. Acad. Sci. USA* **63**, 378–383.
2. John, H. A., Birnstiel, M. L., and Jones, K. W. (1969) RNA-DNA hybrids at the cytological level. *Nature* **223**, 582–587.
3. Alwine J. C., Kemp, D. J., and Stark, G. R. (1977) Method for detection of specific RNAs in agarose gels by transfer to diazobenzyloxymethyl-paper and hybridization with DNA probes. *Proc. Natl. Acad. Sci. USA* **74**, 5350–5354.
4. Parker, R. M. C., Fleetwood-Walker, S. M., Rosie, R., Munro, F. E., and Mitchell, R. (1993) Inhibition by NK_2 but not NK_1 antagonists of carrageenan-induced preprodynorphin mRNA expression in rat dorsal horn lamina I neurons. *Neuropeptides* **25**, 213–222.
5. Rabadan-Diehl, C., Lolait, S. J., and Aguilera, G. (1995) Regulation of pituitary vasopressin V1b receptor mRNA during stress in the rat. *J. Neuroendocrinology* **7**, 903–910.
6. Ciossek, T., Millauer, B., and Ullrich, A. (1995) Identification of alternatively spliced mRNAs encoding varients of MDK1, a novel receptor tyrosine kinase expressed in the murine nervous system. *Oncogene* **10**, 97–108.
7. Rigby, M., Le Bourdelles, B., Heavens, R. P., et al. (1996) The messenger RNAs for the N-methyl-D-aspartate receptor subunits show region-specific expression of different subunit composition in the human brain. *Neuroscience* **73**, 429–447.
8. Gustafson, E. L., Durkin, M. M., Bard, J. A., Zgombick, J., and Branchek, T. A. (1996) A receptor autoradiographic and in situ hybridisation analysis of the distribution of the 5-HT_7 receptor in rat brain. *Br. J. Pharm.* **117**, 657–666.
9. Taketazu, F., Kato, M., Gobl, A., et al. (1994) Enhanced expression of transforming growth factor-beta s and transforming growth factor-beta type II receptor in the synovial tissues of patients with rheumatoid arthritis. *Laboratory Investigation* **70**, 620–630.
10. Harrington, K. A., Augood, S. J., Faull, R. 1., McKenna, P. J., and Emson, P. C. (1995) Dopamine D1 receptor, D2 receptor, proenkephalin A and substance P gene expression in the caudate nucleus of control and schizophrenic tissue: a quantitative cellular in situ hybridisation study. *Brain Research. Mol. Br. Res.* **33**, 333–342.

11. Adcock, I. M., Peters, M., Gelder, C., Shirasaki, H., Brown, C. R., and Barnes, P. J. (1993) Increased tachykinin receptor gene expression in asthmatic lung and its modulation by steroids. *J. Mol. Endocrinology* **11**, 1–7.
12. Kia, H. K., Miquel, M. C., McKernan, R. M., et al. (1995) Localization of 5-HT_3 receptors in the rat spinal cord: immunohistochemistry and in situ hybridization. *Neuroreport.* **6**, 257–261.
13. Noguchi, K., Kawalski, K., Traub, R., Solodkin, A., Iadarola, M. J., and Ruda, M. A. (1991) Dynorphin expression and Fos-like immunoreactivity following inflammation induced hyperalgesia are colocalised in spinal cord neurones. *Mol. Brain Res.* **10**, 227–233.
14. Morales, M. and Bloom, F. E. (1997) The 5-HT_3 receptor is present in different subpopulations of GABAergic neurons in the rat telencephalon. *J. Neuroscience* **17**, 3157–3167.
15. Sambrook, J., Fritsch, E. F., and Maniatis, T. eds. (1989) *Molecular Cloning: A Laboratory Manual.* 2nd edition. Cold Spring Harbour Laboratory, Cold Spring Harbor, New York.
16. Hames, B. D. and Higgins, S. J., eds. (1987) *Nucleic Acid Hybridisation A Practical Approach.* IRL, Oxford University Press, Oxford, UK.
17. Valentino, K. L., Eberwine, J. H., and Barchas, J. D., eds. (1987) *In Situ Hybridisation: Applications to Neurobiology.* Oxford University Press, New York: pp. 57–58.
18. Wilkinson D. G., ed. (1993) *In situ Hybridisation: A Practical Approach.* IRL Press Oxford University Press, New York.
19. Princivalle, A. P., Duncan, J. S., Thom, M., and Bowery, N. G. (2003) GABAB1a, GABAB1b and GABAB2 mRNA variants expression in hippocampus resected from patients with temporal lobe epilepsy. *Neuroscience* **122**, 975–984.
20. Belelli, D., Balcarek, J. M., Hope, A. G., Peters, J. A., Lambert, J. J., and Blackburn, T. P. (1995) Cloning and functional expression of a human 5-hydroxytryptamine type 3As receptor subunit. *Mol. Pharmacol.* **48**, 1054–1062.
21. Parker, R. M. C., Barnes, J. M., Ge, J., Barber, P. C., and Barnes, N. M. (1996) Autoradiographic distribution of [^{3}H]-(s)-zacopride-labelled 5-HT_3 receptors in human brain. *J. Neurol. Sci.* **144**, 119–127.
22. Dopazo, J., Zanders, E., Dragoni, I., Amphlett, G., and Falciani, F. (2001) Methods and approaches in the analysis of gene expression data. *J. Immunol. Methods* **250,** 93–112.
23. Chomczynski, P. and Sacchi, N. (1987) Single-step method of RNA isolation by acid guanidinium thiocyanate-phenol-chloroform extraction. *Anal. Biochem.* **162**, 156–159.
24. Kaupmann, K., Huggel, K., Heid, J., et al. (1997) Expression cloning of $GABA_B$ receptors uncovers similarity to metabotropic glutamate receptors. *Nature* **368**, 239–246.
25. Battaglia, G., Princivalle, A., Forti, F., Lizier, C., and Zeviani, M. (1997) Expression of SMN gene, the spinal muscolar atrophy determining gene, in the mammalian central nervous system. *Hum. Mol. Genet.* **6**, 1961–1971.

26. Barnes, N. M., and Sharp, T. (1999) A review of central 5-HT receptors and their function. *Neuropharmacology* **38**, 1083–1152.
27. Rigby, P. W. T., Dieckmann, M., Rhodes, C., and Berg, P. (1977) Labelling deoxyribonucleic acid to high specific activity in vitro by nick translation with DNA polymerase I. *J. Mol. Biol.* **113**, 237–251.
28. Feinberg, A. P., and Vogelstein, B. (1983) A technique for radiolabelling DNA restriction endonuclease fragments to high specific activity. *Anal. Biochem.* **132**, 6–13.
29. Lewis, M. E., Sherman, T. G., and Watson, S. J. (1985) In Situ Hybridisation histochemistry with synthetic oligonucleotides: strategies and methods. *Peptides* **6 (Suppl. 2)**, 75–87.
30. Emson, P. C. (1993) In-situ hybridisation as a methodological tool for the neuroscientist. *TINS* **16**, 9–16.
31. Mitchell, B. S., Dhami, D., and Schumacher, U. (1992) Review article: In situ hybridisation: a review of methodologies and applications in the biomedical sciences. *Med. Lab. Sci.* **49**, 107–118.
32. Ratcliff, R. C. (1974) Terminal deoxynucleotydyl transferase. In Boyer, P.D., ed., *The Enzymes, 3rd ed., vol. XIV*. Academic, New York: pp. 105–118.
33. Angerer, L. M. and Angerer, R.C. (1992) In situ hybridisation to cellular RNA with radiolabelled RNA probes. In Wilkinson, D. G., ed. *In Situ Hybridization: A Practical Approach* IRL Press, Oxford University Press, Oxford: pp. 15–32.
34. Höltke, H.J., Ankenbauer, W., Mühlegger, K., et al. (1995) The digoxigenin (DIG) system for non-radioactive labelling and detection of nucleic acids—an overview. *Cell. Mol. Biol.* **41**, 883–905.

5

Radioligand-Binding and Molecular-Imaging Techniques for the Quantitative Analysis of Established and Emerging Orphan Receptor Systems

Anthony P. Davenport and Rhoda E. Kuc

1. Introduction

Radioligand binding is widely used to characterize receptors and to determine their anatomical distribution, particularly the superfamily of rhodopsin-like, seven-transmembrane-spanning G protein-coupled receptors (GPCRs). More than 200 receptors that transduce many important physiological processes and are the target for about 50% of all drugs have been identified in this family *(1–6)* . A further approx 160 or so "orphan" GPCRs have been predicted to exist from the human genome and have mRNA sequences characteristic of 7TM GPCRs, but their endogenous ligands await identification. Most of these receptors have been artificially expressed in cell lines linked to a reporter system to identify when a ligand binds to the receptor (*see* Chapter 2). This "reverse pharmacology" approach continues to be used to screen compounds from existing or new combinatorial libraries of biologically active molecules, and has been very successful. More than 45 receptors have been "de-orphanized" or paired with their cognate ligand, with nearly half of these putative endogenous transmitters turning out to be peptides (*see* **Table 1**). The number of pairings continues to increase. It is estimated that about 70 of the remaining orphan receptors could turn out to have a peptidic ligand *(3)*.

1.1. Why Use Ligand Binding to Characterize Receptors?

Molecular techniques, such as *in situ* hybridization and quantitative reverse transcriptase (RT)-polymerase chain reaction (PCR) (*see* Chapters 3 and 4)

From: *Methods in Molecular Biology, vol. 306: Receptor Binding Techniques: Second Edition*
Edited by: A. P. Davenport © Humana Press Inc., Totowa, NJ

Table 1
Orphan Receptors Recently Paired With Their Cognate Peptidic Ligands

Year	Human peptide	Orphan receptor	Ligand source
1995	Nociceptin/Orphanin FQ	ORL-1/OFQ	Porcine brain
1998	Apelin	Apelin/APJ	Bovine stomach
	Orexin A and B (hypercretin 1+2)	Orexin 1 and 2	Rat brain
	Prolactin Releasing Peptide (PRrP)	hGRP-3/GPR10	Bovine hypothalamus
1999	Ghrelin (motilin related peptide)	Ghrelin/GHS-R	Rat stomach
	Melanin concentrating hormone (MCH)	$MCHR_1$	Rat whole brain
	Motilin	GPR38	Peptide library
	Tuberoinfundibular peptide 39 (TIP39)	PTH_2R	Bovine hypothalamus
	Urotensin-II	UT/GPR14/SENR	Peptide library
2000	Neuromedin U-25 (NMU25)	FM3 and FM4	Peptide library
	Neuropeptides FF and AF	NPFF-R1	Bovine brain
2001	Metastin (Kisspeptin-54)	GPR54	Human placenta
	MCH	$MCHR_2$	Peptide library
	Urocortin II and I (stresscopins)	CRF_2	Peptide library
2002	Relaxin	LGR7 and LGR8	Peptide library
	Bovine adrenomedullary peptide 22 (BAM-22)	SNSR3 and SNSR4	Peptide library
	Neuropeptides B and W	GPR7 and GPR8	Peptide library
2003	QRFP43 (P52)	SP9155/GPR103	Genome database
	Relaxin-3/INSL7	GPR135/SALPR	Porcine brain

See **refs.** ***2*** and ***3*** for further information and citations to original papers.

can provide unambiguous evidence for the presence of mRNA encoding a particular novel receptor or receptor subtype in specific cells or tissues *(7–9)*. However, these methods do not provide any information as to whether the protein is actually expressed. Receptors can be identified by their amino acid structure using site-directed antisera (*see* Chapter 8) and functional studies can provide quantitative measurement of affinities. However, ligand binding assays are needed to measure a second key parameter, receptor density (B_{max}), and combined with quantitative autoradiography or phosphor imaging (*see* Chapter 10) can be used to visualize the distribution of receptors within tissue. The selectivity of a receptor may vary depending on posttranslational modifications, other receptors present in tissue to form homodimers and heterodimers, or by the presence of other proteins, such as receptor activity-modifying proteins (RAMPS). Importantly, these may be altered with disease and cannot be predicted using molecular techniques.

1.2. Applications to the Discovery of Novel and Emerging Receptor Systems

Ligand-binding assays provide a powerful tool for identifying and discovering the function of novel orphan receptors recently paired with their cognate ligand *(10–18)*. Following synthesis of a radiolabeled analog of the ligand, expression of a receptor in tissue or cells can be identified by pharmacological criteria that defines receptors; namely, saturable, specific, high-affinity binding. For example, the majority of peptides typically bind with affinity constants in the pM range, with receptor densities in the fM–pM range. The identification of a novel receptor within a tissue or cell can provide an initial clue to possible function and guide the design of functional assays. A characteristic feature of some of the emerging transmitter systems is that the density of receptors can be comparatively low (<10 pmol mg^{-1} protein) and limited to discrete anatomical regions within tissues, which can only be easily detected by quantitative autoradiography. Detection of these receptors can be difficult or impossible using homogenates of whole tissues or organs.

1.3. Applications to Human Disease

The density of receptors in a particular tissue may not be static but can change in response to the concentration of the endogenous agonist. When a GPCR receptor system is exposed to elevated levels of endogenous transmitter (perhaps in response to the progression of disease), the receptor number may undergo a compensatory downregulation, resulting in a reduction in the functional response. Alternatively, low levels of the endogenous ligand may cause a compensatory upregulation. Measurement of such changes by ligand binding can provide initial clues that a particular recently discovered receptor system may be implicated in the disease process, and provide a rationale for further investigation. For example, this strategy was exploited successfully to discover that [^{125}I]-ghrelin receptors are significantly upregulated in human atherosclerosis *(16)*.

1.4. Applications to Animal Models of Human Disease

Changes in binding parameters can be measured following experimental treatment in animal models of human disease *(19–21)* in response to disease processes (*16,22*) or to drugs (*23*). The techniques can be used to characterize transgenic animals in which genes including those encoding receptors have been overexpressed or disrupted to create a global knock-out of a particular receptor system *(21)*. More recently, to avoid developmental or lethal phenotypes, cell-specific knock-outs have been established.

1.5. Saturation-, Competition-, and Kinetic-Binding Assays

This chapter describes the three types of radioligand-binding assays. With saturation assays, tissues or cells are incubated with increasing amounts of radioligands in order to measure the affinity and receptor density, illustrated using the radiolabeled analogue of the peptide ghrelin that has recently been paired with the previously designated orphan GPCR, growth hormone secretagogue receptor (GHS-R; *16*). The application of competition-binding assays is illustrated using ligands selective for the two endothelin (ET) receptor subtypes *(24–28)*, to measure the amount of ET_A and ET_B receptors. Kinetic assays measuring association/dissociation constants are also illustrated with the binding of a peptide, [^{125}I]-PD151242 selective for the endothelin ET_A receptor *(26–28)*. More detailed information about the general theory of receptor binding can be found in **refs.** ***29–31***.

1.6. Nomenclature

In binding assays, it is not possible to distinguish between agonists, and antagonists and both classes of compounds will be referred to as ligands. This chapter will follow the convention that the affinity of a ligand for a receptor, the K_D, is the equilibrium dissociation constant and is a measure of the strength of interaction of a ligand to its receptor *(32)*. The reciprocal of the K_D is the association constant, K_A. By definition, the K_D is the concentration of ligand that will occupy 50% of the receptors. The K_D can be used to calculate the concentration of a radiolabeled ligand needed to occupy a desired proportion of receptors. The fraction of receptors occupied is equal to $L/K_D + L$, where L is the free ligand concentration. For example, a radioligand with a K_D of 0.2 n*M* would occupy 9% of the receptors at a concentration of 0.02 n*M*. The second parameter that can be calculated is the maximum density of receptors, or B_{max}. This is usually corrected using the amount of protein present in the binding assay and expressed as amount of ligand bound/mg protein. The determination of the maximum density of receptors in a particular tissue is unique to ligand-binding and cannot be determined in a functional assay.

2. Materials

2.1. Ligand-Binding Assays

1. Slide-mounted cryostat sections (10–30 μm) from fresh-frozen tissue or cultured cells grown on coverslips.
2. [^{125}I]-labeled radioligands, unlabeled ligands to define nonspecific binding (NSB), and competing ligands for competition assays. Commercial suppliers of radiolabeled ligands include Amersham Biosciences (http://www.amershambioscinces.com) and NEN Life Science Products (http://www.lifesciences.perkinelmer.com).
3. Assay buffer: (*see* **Table 2** and **Note 1** for selecting a suitable buffer), e.g., 0.05 *M* HEPES, 5 m*M* $MgCl_2$, 0.3% bovine serum albumin (BSA) Fraction V, pH 7.4.
4. Ice-cold wash buffer: 0.05 *M* Tris-HCl, pH 7.4 at 4°C.
5. Polypropylene assay tubes (12 × 75 mm, 55.526, e.g., http://www.Sarstedt.co.uk).
6. Slide incubation trays with lids.
7. Metal slide racks.
8. Slide baths.
9. Filter paper to wipe sections from slides.
10. γ-Counter.

2.2. Quantitative Autoradiography and Image Analysis

1. Radiation-sensitive film (Kodak® BioMax MR-1).
2. Autoradiography standards ([^{125}I]-Microscales, RPA 523, Amersham Biosciences).
3. Autoradiography cassettes, boards, and adhesive tape.
4. Kodak D19 developer and Kodak Unifix.
5. Darkroom with safelight for processing films.
6. Image analyzer.

3. Methods

3.1. Saturation Binding

3.1.1. Saturation-Binding Assay

Receptor affinity and density can be readily determined by saturation analysis. In this assay, a number of tissue preparations can be used: partially purified plasma membrane fractions from tissue homogenates, cells transfected with cloned receptors or freshly isolated or cultured cells. The method described as follows uses fresh frozen tissue sections (usually 10–30 μm thick) cut on a microtome and mounted on microscope slides. The latter method has the advantage that radioactivity can be determined in the whole section by wiping the tissue from the slide or by apposing to radiation-sensitive film for autorad-

Table 2
Radioligands, Binding Buffers, and Incubation and Washing Conditions for Established Transmitters Illustrated by Endothelin-1 and Orphan Receptors Recently Paired With Their Cognate Peptidic Ligands

	Radioligand	Binding buffer	NSB	Pre-incubation time (min)	Incubation time (min)	Wash	Reference
A Endothelin	[^{125}I]-ET-1	50 m*M* HEPES, 5 m*M* $MgCl_2$, 0.3% BSA pH 7.4	ET-1 1 μ*M*	15 @ 23°C	120 @ 23°C	3 × 5min 50 m*M* Tris-HCl, pH 7.4 @ 4°C	***25–28***
B Urotensin II	[^{125}I]-Urotensin II (human)	20 m*M* Tris-HCl, 5 m*M* $MgCl_2$, 0.2% BSA, pH 7.4	Urotensin II 10 μ*M*	60 @ 23°C	60 @ 23°C	10 min 50 m*M* Tris-HCl, pH 7.4 @ 4°C	***10***
C Neuropeptide W	[^{125}I]-NPW-23	50 m*M* Tris-HCl, 5 m*M* EDTA, 10 m*M* $MgCl_2$, 0.3% BSA, pH 7.2	NPW-23 1 μ*M*	15 @ 23°C	120 @ 23°C	3 × 5 min 50 m*M* Tris-HCl, pH 7.4 @ 4°C	***13***
D Apelin	[^{125}I]-(Pyr^1)Apelin-13	50 m*M* HEPES, 100 m*M* NaCl, 1 m*M* EDTA, 5 m*M* $MgCl_2$, 10 m*M* KCl, pH 7.4 (7.4)	(Pyr^1) Apelin-13 1 μ*M*	20 @ 23°C 20 m*M* Hepes 1 m*M* EDTA 0.3% BSA Protease inhibitor cocktail pH 7.4	30 @ 23°C	Rapid 50 m*M* Tris-HCl, pH 7.4 @ 4°C	***14***
E Ghrelin	[^{125}I-His^9]-ghrelin	50 m*M* Tris-HCl, 5 m*M* EDTA, 10 m*M* EGTA, AEBSF 1 m*M**, pH 7.2 (7.2)	Hexarelin 1 μ*M*	15 @ 23°C	25 @ 23°C	Rapid 50 m*M* Tris-HCl, pH 7.4 @ 4°C	***16***
F Motilin	[^{125}I]-Motilin	50 m*M* HEPES, 5 m*M* $MgCl_2$, 0.3% BSA, pH 6.9	Motilin 10 μ*M*	15 @ 23°C	120 @ 23°C	3 × 5min 50 m*M* Tris-HCl, pH 7.4 @ 4°C	***17***
G Urocortin-II/III	[^{125}I]-antisauvagine	50 m*M* Tris-HCl, 100 mM NaCl, 10 m*M* $MgCl_2$, 0.3% BSA, pH 7.4	Urocortin I 1 μ*M*	15 @ 23°C	30 @ 23°C	2 × 5 min 50 m*M* Tris-HC1, pH 7.4 @ 4°C	***18***

NSB, nonspecific binding; BSA, bovine serum albumin; EDTA, ethylenediamine tetraacetic acid; EGTA, ethyleneglycol teraacetic acid; AEBSF, aminoethylbenzenesulfonylfluoride.

iography. It avoids lengthy homogenization procedures that can result in degradation of receptor protein and permits the visualization of anatomical regions, which can be important in comparing normal with diseased or control vs experimental tissue.

Saturation assays will be illustrated using the peptide, $[^{125}I]$-ghrelin, in slide-mounted tissue sections of human heart ***(16)***. An initial starting point is to label sections with radioligand concentrations that are selected to span a range at least one order of magnitude above and below the K_D of the ligand, if this is already known (*see* **Note 1**). NSB is determined at each concentration by co-incubation of adjacent sections with a 1000-fold excess over the K_D of an unlabeled ligand that is known to compete at the same receptor. For $[^{125}I]$-ghrelin, the synthetic GHS-R agonist hexarelin is used. The sections are incubated for a defined period of time at a constant temperature in order to ensure equilibrium conditions; these conditions will have been determined previously in association experiments (*see* **Subheading 3.3.1.**). For $[^{125}I]$-ghrelin, a 25-min incubation at 23°C is sufficient to reach equilibrium. Equilibrium is rapidly broken by washing to separate bound from free ligand, and conditions for this should be examined to find an optimal compromise between retention of bound label and a high percentage of specific binding (*see* **Note 2**). Sections are wiped from the slide and the amount of radioactivity bound to the tissue is measured. For homogenates, separation of the bound from the free ligand can be achieved by rapid filtration of the incubation mixture through a filter or by centrifugation.

1. Cut consecutive cryostat sections (typically 10–30 μm) of fresh-frozen tissue and thaw-mount onto gelatin-coated microscope slides. Allow to air dry briefly and store at –70°C until required. Typically, 20 sections (10 Total and 10 NSB) are required for each saturation curve together with a further three sections collected into microcentrifuge tubes to measure protein (*see* **Note 3**).
2. Dilute the stock solution of $[^{125}I]$-ghrelin (5×10^{-8} *M*) to give the highest concentration of 1×10^{-9} *M* by adding 20 μL of stock solution to 980 μL of assay buffer (1 in 50).
3. Using the highest concentration, prepare a serial dilution (500 μL label + 500 μL assay buffer) to give a total of 10 concentrations over the range 1.95×10^{-12} – 1×10^{-9} *M*. Vortex, and use a new pipet tip between each dilution.
4. Remove 250 μL from each of the serial dilutions to determine the NSB, leaving 250 μL to measure total binding (Totals). Add 2.5 μL of 1×10^{-4} *M* unlabeled hexarelin to each NSB tube to give a final concentration of 1×10^{-6} *M* to define the NSB.
5. Count 20 μL aliquots of each Total concentration in a γ-counter, to determine the amount in disintegrations per minute (DPM) of radioligand added for each dilution. These values are required in subsequent analysis of the data (*see* **Subheading 3.1.4.**).

6. In an incubation tray, pre-incubate 20 microscope slides bearing consecutive tissue sections (10 Total, 10 NSB) with 200 μL assay buffer for 15 min at room temperature to remove endogenous ligand and degradative enzymes. Usually, the same buffer can be used during the pre-incubation stage as that used in the subsequent incubation period with the ligand. However, *see* **Note 1** and **Table 2** for choosing alternative pre-incubation buffers containing peptidase inhibitors, which may enhance specific binding for some ligands.
7. Tip off pre-incubation buffer into the tray and replace with 200 μL of each Total or NSB solutions. Cover with a lid to maintain the humidity, and incubate for 25 min at room temperature to reach equilibrium (*see* **Note 1** and **Table 2** for choosing optimum incubation times).
8. Break equilibrium by transferring slides to racks and washing in 400-mL baths containing ice-cold 0.05 *M* Tris-HCl buffer, pH 7.4 at 4°C rapidly (for ghrelin) or by 3 × 5 min washes for other ligands (*see* **Note 2** and **Table 2** for guidance on optimising washing times).
9. Drain and wipe each section from the slide with a filter paper circle, transfer to a counting tube, and count in a γ-counter to measure DPM.
10. For autoradiography following **step 8**, rinse sections once in de-ionized water to remove buffer salts, and dry rapidly in a stream of cold air prior to apposing to radiation sensitive film.

3.1.2. Graphical Analysis

An example of a saturation isotherm for the binding of [^{125}I]-ghrelin to human heart is shown in **Fig. 1.** The total, specific, and nonspecific binding has been plotted against increasing concentrations of the radiolabeled ligand. NSB occurs because the ligand may also bind to other proteins and lipids. This is linear and not saturable. The amount of specific binding observed where the isotherm reaches a plateau is an approximation of the B_{max}. An estimate of the K_D can also be derived from the concentration of [^{125}I]-ghrelin that labels half of the target receptors (50% of the B_{max}). However, a better estimate can be achieved by linearising the data using a Scatchard plot (**Fig. 2**). Here the ratio of bound and free radioligand (B/F) is plotted on the ordinate against the amount of bound radioligand (B) on the abscissa. The slope of the line through the points is equal to the negative reciprocal of the K_D and the intercept of the line with the abscissa is an estimate of B_{max}. This latter parameter is generally standardized against a suitable reference that allows a direct comparison of receptor density in different tissues and in different studies (*see* **Note 3**).

A linear Scatchard plot indicates that the radioligand binds with a single affinity. Under certain circumstances, the Scatchard plot yields a concave line. indicating either that the ligand binds to multiple populations of sites with differing affinities for the radioligand or negative co-operativity. A curvilinear convex line may indicate positive co-operativity. The Hill plot of log [Bound/B_{max}-Bound] vs

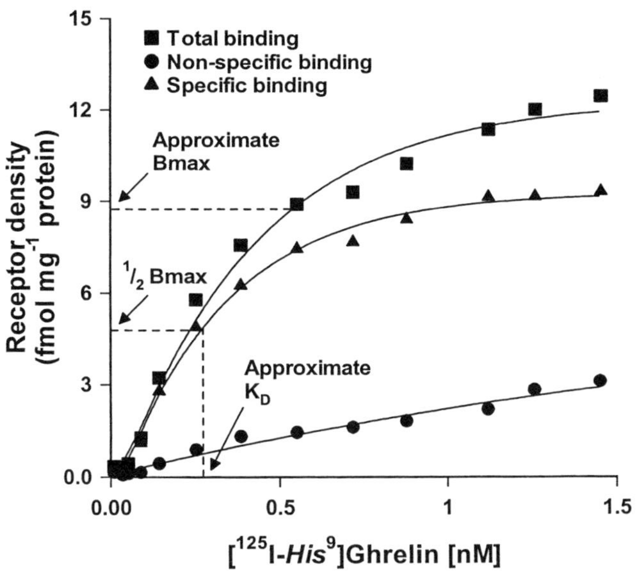

Fig. 1. An example of a saturation isotherm for the binding of increasing concentrations of [^{125}I]-ghrelin to sections of human ventricle. The amount of radioactivity bound to the tissue, measured in DPM by γ-counting, has been converted to fmol of [^{125}I]-ghrelin per mg of protein present in the incubation mixture. Nonspecific binding (NSB), which is linear and not saturable, was subtracted from the total to give the amount of specific binding. The specific binding saturates and an approximate value for the B_{max} for this experiment can be estimated from the ordinate. An approximate value for the K_D can be obtained from the abscissa, corresponding to 50% of the B_{max}.

log [Free] is used to interpret these possible anomalous binding interactions (**Fig. 3**). The Hill coefficient (the slope of the line through the points) will be approximately 1 when the ligand binds to a single population of sites or when the ligand is binding to multiple populations of sites with similar affinity for the radioligand. A value less than 1 may indicate either negative co-operativity or multiple populations of binding sites with differing affinities for the radioligand. In saturation experiments, [^{125}I]-ghrelin binds with Hill slopes close to 1.

3.1.3. Nonlinear Curve Fitting Program

In practice, nonlinear iterative curve-fitting programs are used to calculate final estimates of the binding parameters. Nonlinear regression analysis utilizes equations that can define curves. KELL (Biosoft, http:///www.Biosoft.com) contains Equilibrium Binding Data Analysis (EBDA) and LIGAND programs ***(33,34)***. This suite is recommended because a number of saturation curves can be co-

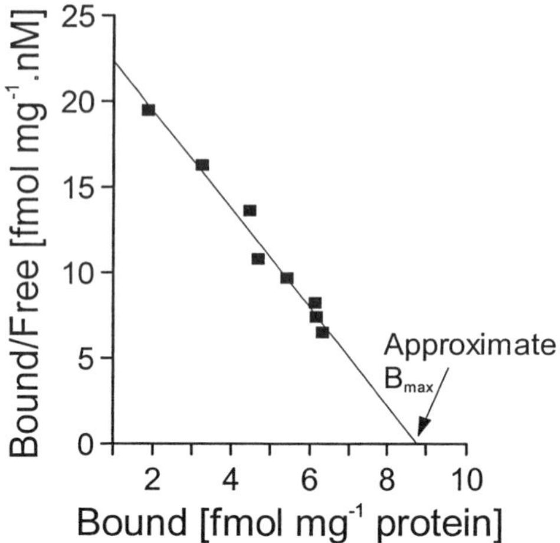

Fig. 2. Scatchard transformation of the data from **Fig. 1.** The slope of the line is equal to the negative reciprocal of the K_D (approx 0.3 n*M*) and the intercept of the line with the abscissa is an estimate of B_{max} (approx 9 fmol/mg protein). These values were then used as initial estimates in the computer program LIGAND, which uses nonlinear curve fitting to calculate the final binding parameters for [^{125}I]-ghrelin by co-analyzing simultaneously binding data from six individuals to give a mean K_D = 0.4 n*M*, and mean B_{max} = 8 fmol mg^1 protein in this tissue ***(16)***.

analyzed simultaneously. EBDA performs the preliminary analysis of both saturation and competition-binding experiments, converting radioactivity in DPM into molar concentrations of ligand. Hill slopes are calculated separately, and a Scatchard transformation is used to provide initial estimates of K_D and B_{max}. LIGAND uses the files created by EBDA, together with initial estimates of the binding parameters, to fit the data to a specified model of the radioligand binding, which may be to one, two, or more sites. A weighted, nonlinear curve-fitting routine is iteratively refined to provide more accurate estimates of the K_D and B_{max} values than those obtained with linear (Scatchard) transformations alone.

A runs test is used to determine whether the data points differ significantly from the fitted curve. A good fit will have points randomly distributed around the fitted line, whereas a run of consecutive points, appearing either above or below the fitted line, may provide evidence for a significant lack of fit. In a saturation assay over a limited concentration range, many radioligands will bind with a single affinity, and a one-site fit is an appropriate model. However, curvilinear Scatchard plots, displayed in the EBDA results (or a significant departure of the data from the fitted curve in the runs test), may suggest that the ligand is binding with different affinities to more than one population of receptors. Therefore, a two- or possibly three-site fit should be tested, and compared

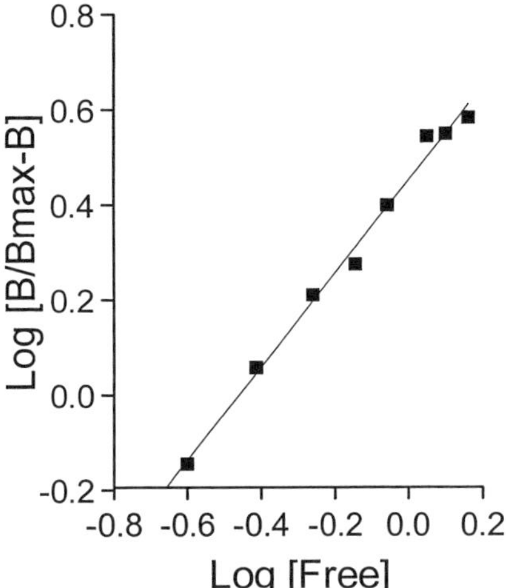

Fig. 3. Hill plot of the data from **Fig. 1.** The slope of the line = the Hill coefficient (nH). For [^{125}I]-ghrelin, the Hill slope (nH) was close to a value of one (0.98), indicating that the radioligand was binding with a single affinity to receptors within the heart tissue, consistent with the linear Scatchard plot.

with the one-site model within the LIGAND program. LIGAND calculates an F value, taking into account the improvement in the fit, that accompanies an increase in the number of parameters to be fitted in the two-site model compared with a one-site fit. The program will indicate which of the two models is statistically the better fit.

3.1.4. Analysis of Saturation-Binding Data: Running EBDA and LIGAND

1. Run the EBDA program, enter assay type (HOT), data type (DPM), specific activity of the label in dpm/pmol (typically 4440000 for a label with specific activity of approx 2000 Ci/mmol), volume of incubation in mL (0.2 mL in above assay), and calculation type (specific bound).
2. For each concentration, enter values (in DPM) for Total and NSB from the γ-counter and press "ESC" to initiate calculations.
3. Check raw data input for accuracy prior to analysis and save data as an EBDA file.
4. Select curve-fitting, and select the model to be fitted (one-site, two-site, and so on) starting with a one-site model. Examine the initial estimates for a one-site model, and if these are reasonable, begin curve-fitting. The program calculates initial estimates of the K_D and B_{max}. Print these results to be used as initial estimates by the Ligand program and create a Ligand file.
5. Run the Ligand program and load the Ligand file created in EBDA. Up to 10 files from separate saturation assays can be loaded and co-analyzed. In this program the following notation is used for fixed parameters:

N— B/F ratio at infinite free concentration and used to calculate NSB.

C—This is a conversion factor used to adjust the amount of protein in different experiments if required, but is not used in the following example.

The following parameters are calculated:

K—This is the association constant of the radioligand ([^{125}I]-ghrelin), K_A, being calculated and is the reciprocal of the K_D value. LIGAND uses K_A instead of K_D.

R—B_{max}, the maximum density of binding sites. This value is corrected using the amount of protein per section(s) mounted on the microscope slide in the incubation volume (*see* **Note 3**).

Start curve-fitting by selecting the number of sites to be fitted (one-site, two-site, and so on) (*see* **Note 4**).

6. Nominate the fixed values: for the NSB, the constant N1 is zero because the individual NSB values have been entered and C1 is set to 1.
7. Set the initial estimates for the floating parameters: K11, the initial estimate obtained from EBDA of the K_D (entered as the reciprocal of the K_D) and R1, the initial estimate from EBDA of the B_{max}.
   The program iteratively refines the initial estimates and, upon convergence, displays the final estimates for the K_D (K11) and B_{max} (R1) together with the standard error (*see* **Note 4**).
8. Re-run EBDA, and under Saturation, select Hill analysis in order to calculate the Hill slope together with the standard error.

3.2. Competition Binding

3.2.1. Competition-Binding Assays

Having determined the K_D of a radiolabeled ligand for a target receptor in a saturation assay, this information can be used to determine the ability of other unlabeled compounds tested over a much wider concentration range (typically 10 p*M*–100 μ*M*), to compete for the binding of a fixed concentration of the labeled ligand. Competition-binding experiments can also be used to determine the selectivity (if any) of a particular ligand for ET_A or ET_B receptors, and thus allow determination of the density and proportion of each subtype in the tissue. Competition curves are obtained by plotting specific binding as a percentage of total binding (binding in the absence of competitor) against the log concentration of the competing ligand. Both ET receptor subtypes are present in left ventricle of the human heart in a ratio of about 60% ET_A:40% ET_B, allowing ligands to be characterized against both receptors in the same tissue. In **Fig. 4**, [^{125}I]-ET-1 has been used to label all of the ET receptors in this tissue ***(25–28)*** because the peptide has equal affinity for both the ET_A and ET_B subtype. A steep competition curve is usually indicative of binding to a single population of receptors. However, increasing concentrations of unlabeled PD151242 inhibited the binding of [^{125}I]-ET-1 biphasically. A shallow

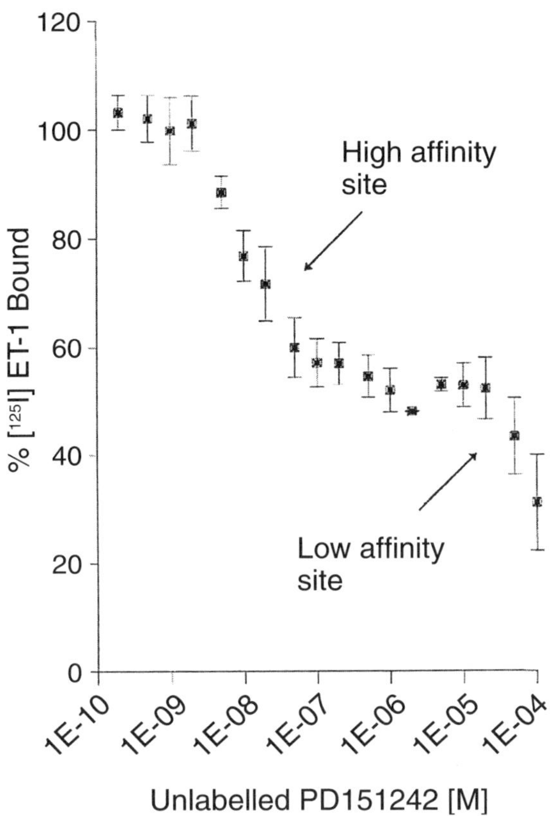

Fig. 4. Competition-binding curve for the inhibition of a fixed concentration of [^{125}I]-ET-1 (0.1 n*M*) binding to endothelin (ET) receptors by increasing concentrations of unlabeled PD151242 in sections of human ventricle. Over the concentration range tested, PD151242 competed in a biphasic manner and a two-site fit was preferred to a one-site or three-site model using LIGAND. The high affinity site corresponded to the endothelin ET_A receptor ($K_D = 7.2 + 2.8$ n*M*), the low affinity site to the ET_B receptor, $K_D = 104 + 23$ μ*M*. Each value represents the mean ± SEM of three individuals.

curve or a curve with clear inflection points is indicative of multiple populations of binding sites. The partial F test and the runs test can again be used to differentiate single and multiple populations of binding sites. In this case, a two-site fit was preferred, consistent with PD151242 binding with high affinity to the ET_A site but low, micromolar affinity to the ET_B receptors, giving about 15,000-fold selectivity for the ET_A receptor.

The plot of log (%B/[100–%B]) against log [L*], where %B is the percentage of radioligand bound and [L*] is the concentration of the competing ligand, gives a line through the points with slope equal to the pseudo Hill coefficient. This approximates unity for a one-site fit and is less than unity if negative co-operativity is indicated or if two or more populations of receptors with differing affinities for the competing ligand are present. However, binding constants for multiple receptor populations are difficult to determine accurately by graphical means. EBDA uses an equation that allows interpretation of a heterogeneous population of binding sites and provides estimates of the pseudo Hill coefficient and receptor affinity. LIGAND is then used to test for one, two, or more binding sites as described for the analysis of saturation data.

1. Cut 25 consecutive cryostat sections (typically 10–30 µm) of fresh-frozen tissue per competition assay. Two sections are used to measure total binding and 2 sections to measure NSB, leaving 21 sections for construction of the competition curve by the unlabeled ligand. A further three sections are collected into microcentrifuge tube to measure protein.
2. Dilute the stock solution of radiolabeled [^{125}I]-ET-1 (5×10^{-8} *M*) in assay buffer to give a concentration of 1.1×10^{-10} *M* (1:454.5 dilution equivalent to 13.2 µL in 6 mL of assay buffer).
3. Prepare a stock concentration (1×10^{-3} *M*) of the competing ligand. Dilute this in assay buffer to give three further stock solution solutions (1×10^{-4}, 2×10^{-4} and 5×10^{-4} *M*) as shown in **Table 3**. From each of three solutions, prepare a series of dilutions to give a concentration range $1 \text{x} 10^{-3} - 2 \times 10^{-10}$ *M*. Vortex, and use a new pipet tip between each dilution.
4. Pipet 225 µL of [^{125}I]-ET-1 into 2 tubes labeled Total, 2 labeled NSB, and a further 21 labeled with each final concentration of competing ligand (1×10^{-4} – 2×10^{-11} *M*).
5. Pipet 25 µL of assay buffer into the two Total tubes to give a final concentration of [^{125}I]-ET of 1×10^{-10} *M*. Vortex.
6. Pipet 25 µL of 1×10^{-5} *M* unlabeled ET-1 into the two NSB tubes to give a final concentration of 1×10^{-6} *M*. Vortex.
7. Pipet 25 µL of each prepared concentration of competing ligand into the appropriately labelled tube. Vortex, and use a new pipet tip between each solution.
8. In incubation trays, pre-incubate 25 microscope slides bearing consecutive tissue sections (2 Total, 2 NSB, 21 competing ligand) with 200 µL assay buffer for 15 min at room temperature to remove endogenous ligand and degradative enzymes.
9. Tip off pre-incubation buffer into tray and replace with 200 µL of Total, NSB, or competing ligand dilution. Cover with lid to maintain humidity and incubate for 120 min at room temperature to reach equilibrium.
10. Break equilibrium by transferring slides to racks and washing in 400-mL baths containing ice-cold 0.05 *M* Tris-HCl, pH 7.4 at 4°C (3×5 min).
11. Drain and wipe each section from the slide with a filter paper circle, transfer to a counting tube, and count in a γ-counter to measure DPM.

Table 3
Competition Binding Assay

a	Label µL	Unlabeled ligand µL	Competing ligand µL	Buffer µL
Total (×2)	225	–	–	25
NSB (×2)	225	25	–	–
Competing ligand (×21 Concentrations)	225	–	25	–

b	Series 1	Series 2	Series 3
Stock (1E-3M) µL	10	20	50
Buffer µL	90	80	50
Target Concentration	⇓	⇓	⇓
[*M*]	1E-4	2E-4	5E-4

c	Series 1 1E-4	Series 2 2E-4	Series 3 5E-4
Serial dilution			
10 µL + 90 µL buffer	1E-5	2E-5	5E-5
	1E-6	2E-6	5E-6
	1E-7	2E-7	5E-7
	1E-8	2E-8	5E-8
	1E-9	2E-9	5E-9
		2E-10	5E-10

d Serial Dilution 10 µL + 90 µL buffer	From serial dilutions	Final concentration
1	1E-3 (from stock)	1E-4
2	5E-4	5E-5
3	2E-4	2E-5
4	1E-4	1E-5
5	5E-5	5E-6
6	2E-5	2E-6
7	1E-5	1E-6
8	5E-6	5E-7
9	2E-6	2E-7
10	1E-6	1E-7
11	5E-7	5E-8
12	2E-7	2E-8
13	1E-7	1E-8
14	5E-8	5E-9
15	2E-8	2E-9
16	1E-8	1E-9
17	5E-9	5E-10
18	2E-9	2E-10
19	1E-9	1E-10
20	5E-10	5E-11
21	2E-10	2E-11

12. For autoradiography following **steps 1–10**, rinse sections once in de-ionized water to remove buffer salts, and dry rapidly in a stream of cold air prior to apposing to radiation sensitive film.

3.2.2. Analysis of Competition-Binding Data: Running EBDA and LIGAND

1. Run the EBDA program, enter assay type (Competition), data type (DPM), specific activity of the label in dpm/pmol (typically 4440000 for a label with specific activity of approx 2000 Ci/mmol), volume of incubation in ml (0.2 mL in above assay), and calculation type (specific bound).
2. Enter Total and NSB in DPM (common to all subsequent data points).
3. Enter each concentration (2×10^{-11} *M* is entered as 2E-11, and so on) and its corresponding counts in DPM from the γ-counter.
4. Check raw data input for accuracy and save data as an EBDA file to be used later by Ligand.
5. Select curve-fitting and select the model to be fitted (one-site, two-site, and so on), starting with a one-site model, and start curve-fitting. In the above assay, unlabeled PD151242 is expected to compete for the binding of [^{125}I]-ET-1 biphasically with a high affinity (n*M*) for ET_A and low affinity (μM) at ET_B receptors. A one-site model is unlikely to be fitted so a two-site model should be chosen. Examine the initial estimates for a two-site model, and if these are reasonable, begin curve-fitting. The program calculates initial estimates of the high- and low-affinity K_D values together with their corresponding receptor densities (B_{max}). Print these results to be used as initial estimates by the Ligand program and create a Ligand file.
6. Run the Ligand program and load the data file created in EBDA. Up to 10 files from separate assays can be loaded and co-analyzed. In this program the following notation is used:
   K—This is the association constant K_A which is the reciprocal of the K_D value. LIGAND uses K_A instead of K_D.
   R—B_{max}, the maximum density of binding sites. This value is corrected using the amount of protein per section(s) mounted on the microscope slide in the incubation volume (*see* **Note 3**).
   N—B/F ratio at infinite free concentration and used to calculate NSB.
   C—this is a conversion factor used to adjust the amount of protein in different experiments if required, but not used in the following example.

3.2.2.1. One-Site Fit

1. Start curve-fitting by selecting a one-site fit.
2. Set the constant parameters: N2 C1 K11.
   N2—NSB of competing ligand (PD151242) set to 0.
   C1—set to 1.
   K11— refers to the ligand binding to site 1, which for [^{125}I]-ET-1 is the reciprocal of the K_D.

3. Set the initial estimates of the floating parameters: K21, the estimate obtained from EBDA of the K_D for PD151242 (entered as the reciprocal of the K_D), and R1, the initial estimate from EBDA of the B_{max}.
4. The program iteratively refines the initial estimates and upon convergence displays the final estimates for the K_D (K21) and B_{max} (R1) together with the standard error.

3.2.2.2. Two-Site Fit

1. Start curve-fitting by selecting a two-site fit.
2. Set the constant parameters: N1 C1 K11 K12 N2.
   K11—refers to the radioligand ($[^{125}I]$- ET-1) binding to site 1 (ET_A).
   K12—refers to the radioligand ($[^{125}I]$-ET-1) binding to site 2 (ET_B).
   Because $[^{125}I]$-ET-1 has equal affinity for both ET_A and ET_B receptors, these two values will be the same, and for $[^{125}I]$-ET-1 the reciprocal of the K_D is entered.
   N1—NSB for ligand 1 and is set to 0.
   N2—NSB of competing ligand (PD151242) set to 0.
   C1—is a conversion factor that can be applied but is not used in this example and is set to 1.
3. Set the initial estimates of the floating parameters: K21 and K22.
   K21—refers to ligand 2 (unlabeled PD151242) binding to site 1 (ET_A).
   K22—refers to ligand 2 (unlabeled PD151242) binding to site 2 (ET_B).
   Both parameters are obtained from EBDA as initial estimates for the K_D for PD151242 (entered as the reciprocal of the K_D).
   R1 and R2—are the initial estimates from EBDA of the B_{max} values for the two sites.
4. The program iteratively refines the initial estimates and, upon convergence, displays the final estimates for the K_D (K21) of PD151242 binding at the high-affinity, ET_A site and K21, the low-affinity, ET_B site. The density of ET_A (R1) and ET_B receptors (R2) is also given (*see* **Note 4**).

3.3. Binding Kinetics

Kinetic experiments determine the time course of ligand association and dissociation (**Fig. 5**). In association studies, sections are labeled with a fixed concentration of radioligand for increasing time periods. NSB is defined at each time point using a high concentration of unlabeled ligand. The plot of $\ln(B_{eq}/(B_{eq}-B_t)$ against time, where B_{eq} is the amount of ligand bound at equilibrium and B_t is the amount of ligand bound at time t, should yield a straight line through the points with slope equal to K_{obs}.

When equilibrium is reached, dissociation of the radioligand from the receptors is achieved either by incubation of tissue sections with a high concentration of unlabeled competitor or by infinite dilution of the labeled sections by immersion in a large volume of buffer. The plot of $\ln(B_t/B_0)$ against time, where B_0 is binding at time 0, should be linear with the slope equal to the dissociation rate

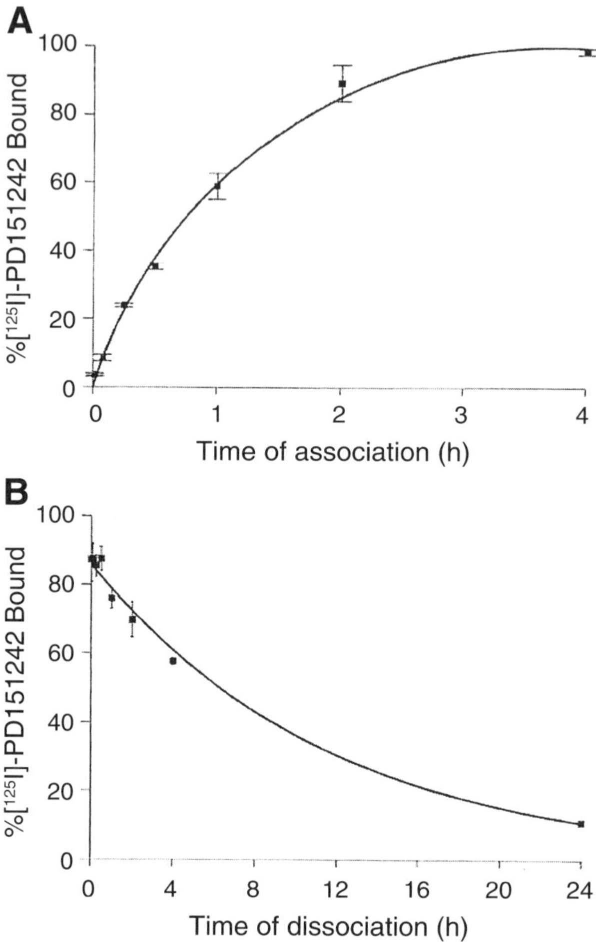

Fig. 5. (**A**) Association curve for a fixed concentration of [^{125}I]-PD151242 binding to sections of human ventricle. The curve plateaus after 2 h, indicating that equilibrium has been reached for this ligand. The calculated observed association rate constant (K_{obs}) was 0.0172 min^{-1}. (**B**) Time course for the dissociation of [^{125}I]-PD151242 initiated by washing the sections in a large volume of buffer. The calculated dissociation rate constant (K – 1) was 0.00144 min^{-1}.

constant (K-1, K21 or K_{off}). The association (K1, K12 or K_{on}) and dissociation rate constants are described by the following relationship: K1 = (K_{obs}–K-1)/[L], where [L] is the concentration of free ligand. Kinetic experiments provide an additional means of calculating the equilibrium dissociation constant, K_D = K-1/K1, and this should be comparable to K_D values determined by saturation analysis (*see* **Subheading 3.1.**). Although the above linear transformations give estimates of association and dissociation constants, the use of computer-based, nonlinear curve-fitting, such as KINETIC in the KELL suite of program (*see* **Subheading 3.1.3.**), are recommended for the analysis kinetic data.

3.3.1. Association-Binding Assays

1. Cut 16 consecutive cryostat sections (typically 10–30 μm) of fresh-frozen tissue per association assay. One section is used to measure total binding, and one section for the NSB at each of eight time points.
2. Dilute the stock solution of radiolabeled [^{125}I]-PD151242 (5×10^{-8} *M*, ***26–28***) to give a concentration of 1×10^{-10} *M* (corrected for decay, if any) in 4 mL of assay buffer (count a 200 μL aliquot in order to determine the total counts added, in DPM, to each slide for subsequent analysis). Remove 2 mL to define Total binding and add unlabeled peptide to a final concentration of 1×10^{-6} *M*, to define the NSB. Vortex.
3. Pre-incubate 16 microscope slides bearing consecutive sections with 200 μL assay buffer for 15 min at room temperature.
4. Tip off pre-incubation buffer into tray and replace with 200 μL of Total (eight slides) or NSB (eight slides). Transfer slides at intervals of 1, 2, 5, 15, 30, 60, 120, and 240 min into racks and wash in 400-mL baths containing ice-cold 0.05 *M* Tris-HCl, pH 7.4 at 4°C (3 × 5 min).
5. Drain and wipe each section from the slide with a filter paper circle, transfer to a counting tube and count in a γ-counter to measure DPM.

3.3.2. Analysis of Association-Binding Data

1. Run the Kinetic program, enter assay type (Association), data type (DPM), specific activity of the label in dpm/pmol (typically 4440000 for a label with specific activity of approx 2000 Ci/mmol), volume of incubation in mL (0.2 mL in above assay), and calculation type (specific bound).
2. Enter Time of sample, and Total and NSB in DPM. Enter the amount of radioactivity added per 200 μL (DPM), which will be the same for each section.
3. Check raw data input for accuracy, view initial estimates, and save data as a Kinetic file.
4. Start curve-fitting. Select the model to be fitted (one-site, two-site, and so on) starting with a one-site model. Commence curve-fitting to refine the model estimates and if acceptable save the result in memory. The observed association rate constant (K_{obs}) is displayed as the exponent together with the error (units min^{-1}).
5. Repeat using a two-site model. In the above example, a second site could not be fitted indicating that a one-site fit was an appropriate model for [^{125}I]-PD151242 binding (**Fig. 5A**).

3.3.3. Dissociation-Binding Assays

1. Cut 16 consecutive cryostat sections (typically 10–30 μm) of fresh-frozen tissue per association assay. One section is used to measure total binding and one section for the NSB at each of eight time points.

2. Dilute the stock solution of radiolabel, [^{125}I]-PD151242 (5×10^{-8} *M*), to give a concentration of 1×10^{-10} *M* (corrected for decay, if any) in 4 mL of assay buffer (count a 200 µL aliquot in order to determine the total counts added, in DPM, to each slide for subsequent analysis). Remove 2 ml to define Total binding and add unlabeled peptide to a final concentration of 1×10^{-6} *M*, to define the NSB. Vortex.
3. Pre-incubate 16 microscope slides bearing consecutive sections with 200 µL assay buffer for 15 min at room temperature.
4. Tip off pre-incubation buffer into tray and replace with 200 µL of Total (eight slides) or NSB (eight slides). Incubate for 2 h to reach equilibrium.
5. Starting with the last pair of slides, transfer slides at intervals of 1, 5, 60, 120, 240, and 1500 min into racks and wash in 400-mL baths containing ice-cold 0.05 *M* Tris-HCl, pH 7.4 at 4°C (3×5 min).
6. Drain and wipe each section from the slide with a filter paper circle, transfer to a counting tube, and count in a γ-counter to measure DPM (**Fig. 5B**).

3.3.4. Analysis of Dissociation-Binding Data

1. Run the Kinetic program, enter assay type (Dissociation), data type (DPM), specific activity of the label in dpm/pmol (typically 4440000 for a label with specific activity of approx 2000 Ci/mmol), volume of incubation in mL (0.2 mL in above assay), and calculation type (specific bound).
2. Enter Time of sample, and Total and NSB in DPM. Enter the amount of radioactivity added per 200 µL (DPM), which will be the same for each section.
3. Check raw data input for accuracy, view initial estimates, and save data as a Kinetic file.
4. Start curve-fitting. Select the model to be fitted (one-site, two-site, and so on) starting with a one-site model. Commence curve-fitting to refine the model estimates and if acceptable save the result in memory. The dissociation rate constant (K–1) is displayed as the exponent, together with the error.
5. Repeat using a two-site model. In the above example, a second site could not be fitted indicating a one-site fit was an appropriate model for [^{125}I]-PD151242 binding. Where a two-site model can be fitted, a partial F test can be used to determine which fit is preferred (**Fig. 5B**).

3.4. Autoradiography

Autoradiography can be used to visualize the localization of receptors in discrete regions of tissue and to confirm that the radioligand binds to cells that would be expected to express the target receptor. The principle of the technique is that the spatial distribution of radiolabeled ligands can be detected by the blackening of radiation-sensitive film apposed directly to the section containing radiolabeled tissue. Slide-mounted tissue sections or cover slips containing cultured cells are used. Sections are pre-incubated in buffer, labeled with the radioligand in the absence or presence of selective competing ligands, washed in buffer, dipped in water to remove buffer salts, and dried rapidly under a stream of cool dry air. The K_D (determined by saturation analysis) is used to calculate the concentration of the radioligand required to label a fixed proportion of receptors in the tissue. Using a concentration calculated to label 10% of the receptors results in a high ratio of Total to NSB. At the end of the assay, sections are apposed to radiation sensitive film for macro-autoradiography and exposed in the dark for a period that will be determined by the type of isotope used, the density of binding sites in the tissue section, and by the tissue section thickness (**Figs. 6** and **7**). The films are then developed, fixed and viewed. When higher resolution is required, sections can be apposed to cover slips coated in nuclear emulsion. Developed silver grains represent the location of the bound radioligand in the underlying tissue when viewed using a microscope equipped with darkfield illumination.

3.4.1. Quantitative Autoradiography Assays

Quantitative autoradiography can be used to analyze a number of different types of experiment. Sections can be apposed to radiation sensitive film following either saturation (*see* **Subheading 3.1.**) or competition-binding (*see* **Subheading 3.2.**) assays. Sections can also be incubated with a fixed concentration of ligand, for example [^{125}I]-ET-1. Adjacent sections are incubated in the presence of a fixed concentration of an unlabeled ET_B ligand to block [^{125}I]-ET-1 binding to this subtype, thus delineating the ET_A receptors. A third section is incubated with a fixed concentration of an unlabeled ET_A ligand to define the ET_B receptors. A fourth section is used to measure NSB ([^{125}I]-ET-1 + 1 μ*M* unlabeled ET-1). For example, for the ET_A selective ligand, BQ123, the K_D at the ET_A receptor = 7×10^{-10} *M* and at the ET_B = 2.4×10^{-5} *M*. Using these K_D values, BQ123 at a concentration of 200 n*M* is calculated to block >99% of [^{125}I]-ET-1 binding to ET_A receptors but <1% of the ET_B. Similarly, for BQ3020, the K_D at the ET_B receptor = 1.4×10^{-9} *M* and at the ET_A= 2×10^{-6} *M*. BQ3020 at a concentration of 200 n*M* can be calculated to occupy >99% of ET_B but <9% ET_A receptors (***8***). Once the appropriate binding assay has been completed and the tissue sections dried, continue as follows.

1. Mount microscope slides bearing tissue sections onto card, together with a microscope slide bearing calibrated radioactive standards ([^{125}I]-Microscales, *see* **ref. *35*** and **Note 5**) in a light tight X-ray cassette. In a darkroom with an appropriate safelight, appose to a single coated radiation sensitive film (Kodak BioMax MR-1) and leave for 2–5 d.
2. In a darkroom with a safelight, monitor the development of autoradiograms for up to 5 min in D19 developer; rinse for 30 s in de-ionized water to stop development. Fix for 30 min in Kodak Unifix (*see* **Note 6**).

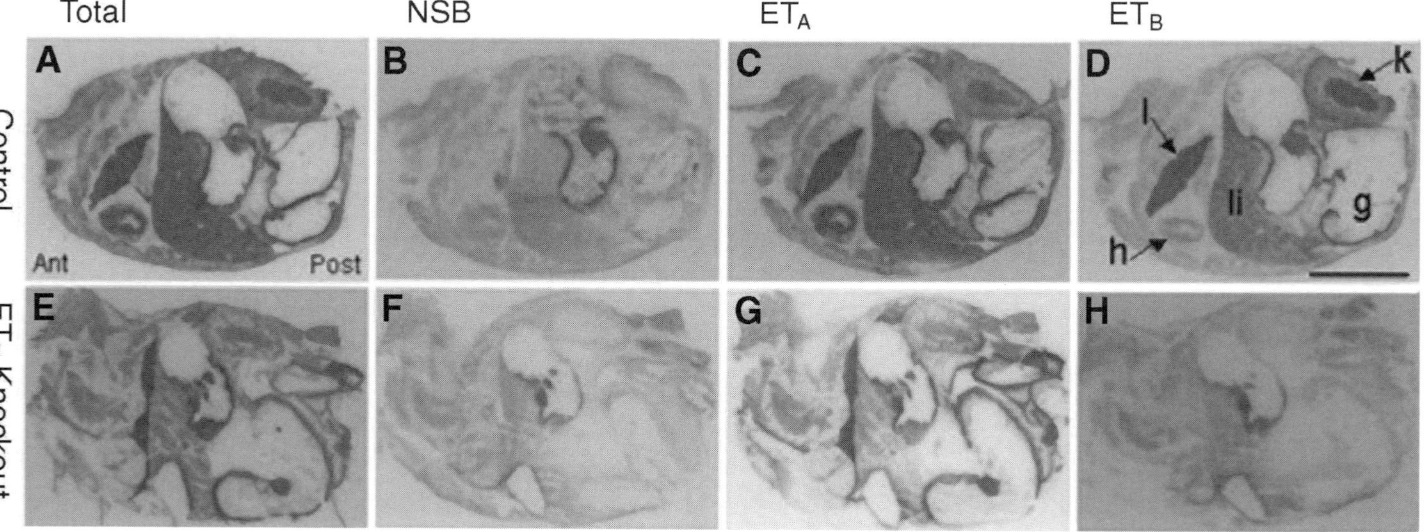

Fig. 6. An example of the phenotypic characterization of deleting a gene for a G protein-coupled (GPC) receptor in mice. Homozygote ET_B knockout mice are viable at birth, and can survive for up to eight weeks but display aganglionic megacolon as a result of absence of ganglion neurons together with a pigmentary disorder in their coats. Autoradiographical localization of endothelin (ET) receptors in consecutive longitudinal cryostat sections cut through the midline of the torso of male control (+\+, upper panel) or homozygous ET_B knock-out mice (–\–, lower panel). To visualize all ET receptors (**A** and **E**, Total), sections were incubated with 0.1 n*M* [^{125}I]-ET-1 alone. Adjacent sections were incubated with the label plus 1 µ*M* ET-1 to define the nonspecific binding (NSB) (**B** and **F**). In **C** and **G**, a concentration of ET_B ligand, BQ3020 (200 n*M*) was calculated (*see* **Subheading 3.4.1.**) to selectively block [^{125}I]-ET-1 binding to ET_B receptors to reveal the ET_A distribution. In **D** and **H**, the concentration of an ET_A selective ligand, BQ123 (200 n*M*), was calculated to bock [^{125}I]-ET-1 binding to this subtype to visualize the ET_B distribution. In the control mouse (**D**), high densities of ET_B binding was detected in kidney and lungs, with lower levels in the liver. In the ET_B knock-out, binding could not be detected, as expected, above the NSB, confirming the targeted disruption of the ET_B gene in all tissues examined. (Ant, anterior; Post, posterior; g, gut; h, heart; k, kidney; l, lung; li, liver. Scale bar = 1 cm).

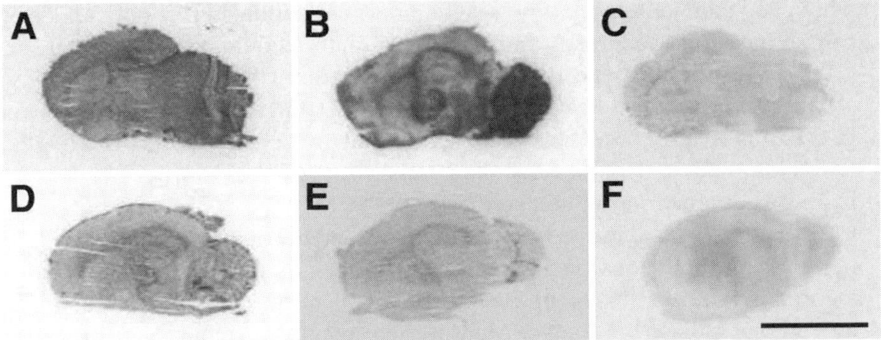

Fig. 7. An example of visualizing and measuring changes in receptors by autoradiography and competition-binding assays. Autoradiographical images of ET_A **(A)**, ET_B **(B)** receptors and **(C)** non-specific binding (NSB) in longitudinal sections of control (+/+) mouse brain were obtained as described in the legend to **Fig. 6**. In ET_B knock-out mice (–/–), a similar distribution but lower density of ET_A receptors were visualised **(D)** but specific binding to ET_B receptors **(E)** could not be detected compared with the NSB **(F)**. (Scale bar = 5 mm). In control brains using competition binding assays, unlabelled BQ3020 competed for [^{125}I]-ET-1 biphasically, with a more abundant high affinity site corresponding to the ET_B receptors and a low affinity, micromolar ET_A site, giving a selectivity for BQ3020 of over 200-fold. In the knock-out mice, however, BQ3020 did not compete for the high affinity ET_B site in agreement with the autoradiographical data. There was no change in the affinity for the ET_A receptor compared with control but ET_A receptor density was significantly reduced by 45% (values are means ± SEM. ET, endothelin; ND, binding not detected above NSB).

Binding Parameters for ET Receptor Subtypes in Mouse Brain

	n	K_D μ*M* (ET_A)	B_{MAX} (ET_A)	K_D n*M* (ET_B)	B_{MAX} (ET_B)
Control	6	2.1 ± 0.8	32.7 ± 6.3	9.3 ± 5.4	48.7 ± 5.8
Knock-out	7	1.9 ± 0.7	18.2 ± 1.9*	ND	ND

*Student's unpaired *t* test, $p < 0.05$

3.4.2. Computer-Assisted Image Analysis

Analyze the resulting autoradiograms by measuring diffuse integrated optical density using a computer-assisted image analysis system (*see* **Note 7**).

1. Calibrate the image analyzer for densitometry. Autoradiograms are illuminated by reflected white light and the image captured by the videoscanner equipped with a zoom lens. Alter the zoom lens mounted on the videoscanner to produce a measuring field appropriate to the size of the autoradiographical image to be analyzed. Set the shading corrector to give an image, which appears uniformly

white, and compensate for any variation in illumination so that optical densities can be measured accurately throughout the autoradiogram. Set the white level (100% transmission) and the scanner dark current (the current flowing in the scanner in the absence of a signal), which would otherwise contribute to the grey image. Finally, calibrate the system against neutral density filters to convert grey levels into optical densities and the number of pixels per unit area is calculated by means of a measuring box (*see* **Note 8**).

2. Construct a standard curve from calibrated standards (*see* **Note 5**) for each film to relate optical densities to known amounts of radioactivity. Detect the autoradiographical image of each standard in turn, using the cursor to draw around and isolate the image. Measure the integrated optical density for each standard together with the area. Enter the amount of radioactivity for each standard, measured by γ-counting in DPM (corrected for the efficiency of the counter and decay, if any), which is divided by the area to calculate radioactivity in dpm/mm^2. The specific activity of the label can be used to convert these values into amol/mm^2. Generate the natural log plot of optical density versus radioactivity to give a linear relationship.
3. Measure the density of ET receptors by digitising each autoradiographical image of the tissue sections. Delineate regions of interest (or use other binary masks such as a circle) from the resulting grey image by using a cursor to draw around a defined anatomical region.
4. When all measurements have been made for a particular section, increase the threshold for detecting the autoradiogram to produce a template that can be used to align the autoradiographical image of the NSB section. Subtract the second image from the first to measure the amount of specific binding. Convert the resulting optical densities to the amount of specifically bound radioligand either in dpm/mm^2 for saturation/competition assays for analysis by EBDA and Ligand or in amol/mm^2 when comparing fixed concentration of ligand by interpolation from the standards curve.

4. Notes

1. In saturation experiments, the amount of radioligand added is increased while maintaining a constant specific activity of the radioligand. In practice, because of cost and hazards of handling high levels of radioacti1vity it is usually not possible to achieve concentrations or [^{125}I]-labeled ligand above 10 n*M* and saturation assays are limited to ligands with affinities in the subnanomolar range. An alternative is to use a constant concentration of radioligand, and the specific activity of the radioligand is decreased by the addition of unlabeled ligand (*see* **Subheading 3.1.**). The assay buffers given in **Table 2** have been optimised for the individual ligands. For a new ligand for which the binding conditions have not been established, it is recommended that a series of buffers should be tested using those given in **Table 2** as a guide, starting with the simplest, buffer A containing HEPES, then buffer B, based on Tris. Additional modifications to buffer include adding ethylenediamine tetraacetic acid (EDTA) (C), higher lev-

els of salts (D), and protease inhibitors (E). Varying the pH can also improve the level of specific binding (F) or increasing levels of salts (G).

NSB is defined by co-incubating adjacent sections with an excess of the unlabeled ligand. For ligands such as peptides that are expected to bind with affinities in the p*M* range, a concentration of 1 μ*M* is sufficient. Using a higher concentration (10 μ*M*) can sometimes reduce further the NSB. Theoretically, ligands reach equilibrium after 5 × *t*1/2 rate for association. For example, if the *t*1/2 is 10 min, equilibrium will be reached after 50 min. Provided the ligand is not metabolized during the incubation period, a useful starting time for incubation is 2 h.

2. The assay is terminated by rapidly breaking the equilibrium by washing slide in a large excess of ice-cold buffer. Optimum washing times from three 1-s dips (for rapidly dissociating ligands) to three five minutes washes should be determined empirically.
3. The B_{max} value in the LIGAND printout is given in mol/L. To convert to pmol of ligand bound per mg protein, divide R1 (if one-site or R1, and R2 if two-site fit, and so on) by the amount of protein measured in representative sections in mg/L.
4. As a guide over the concentration range used in the above saturation assay, [^{125}I]-ghrelin would be expected to bind with similar subnanomolar affinity to GHS receptors in human tissue and a one-site, monophasic model is anticipated. To check, LIGAND should be re-run by fitting a two-site model. In many cases, the program will be unable to do this, leading to an error message of overflow or ill conditioning. Where a two-site fit is obtained, inspection of the resulting F test, which compares the two fits, will show whether a one- or two-site is statistically a better fit.
5. Commercial standards (activity range 1.2–646 nCi/mg) consist of layers containing radioactivity incorporated at the molecular level in a methacrylate copolymer, separated by inert colored layers. Cut strips are expanded on water at 60°C and brushed flat onto gelatin-subbed slides to remove creases. Representative sections are subdivided into the individual activity levels and counted in a γ-counter, to measure the amount of radioactivity in DPM, with correction for the efficiency of the counter. Because standards are designed to be used for up to a year, the amount of radioactivity should be corrected for decay from the time of counting the standards to the mid-time point between apposing and developing the film.
6. Optimum development can usually be assessed visually under safelight. However, if autoradiograms are too dark or too light, sections can be re-apposed to film.
7. A range of image analyzers are available, but in order to carry out accurate densitometry, the machine should be equipped with a shading corrector and be able to digitize images into an array of at least 500 × 500 picture points with a minimum of 256 grey levels. The image analyzer should also be able to subtract stored images. The details of operation of different commercial systems vary considerably, but the major procedures are similar.
8. This should be repeated each time the conditions are changed, such as when the magnification is altered or a different film is used.

Acknowledgments

Supported by grants from the British Heart Foundation.

Reference

1. Wise, A., Jupe, S. C. and Rees, S. (2004) The identification of ligands at orphan G-protien coupled receptors. *Annu. Rev. Pharmacol. Toxicol.* **44,** 43–66.
2. Katugampola, S., and Davenport, A. P. (2003) Emerging roles for orphan G-protein-coupled receptors in the cardiovascular system. *Trends Pharmacol. Sci.* **24,** 30–35.
3. Davenport, A. P. (2003) Peptide and trace amine orphan receptors: prospects for new therapeutic targets. *Curr. Opin Pharmacol.* **3,** 127–134.
4. Davenport, A. P., and Macphee, C. H. (2003) Translating the human genome: renaissance of cardiovascular receptor pharmacology. *Curr. Opin. Pharmacol.* **3,** 111–113.
5. Lee, D. K., George, S. R., and O'Dowd, B. F. (2003) Continued discovery of ligands for G protein-coupled receptors. *Life Sci.* **74,** 293–297.
6. Katugampola, S. D., Maguire, J. J., Kuc, R. E., Wiley, K. E., and Davenport, A. P. (2002) Discovery of recently adopted orphan receptors for apelin, urotensin II, and ghrelin identified using novel radioligands and functional role in the human cardiovascular system. *Can. J. Physiol. Pharmacol.* **80,** 369–374.
7. Davenport, A. P., O'Reilly, G., Molenaar, P., Maguire, J. J. Kuc, R. E., Sharkey, A., et al. (1993) Human endothelin receptors characterised using reverse transcriptase-polymerase chain reaction, in situ hybridization and sub-type selective ligands BQ123 and BQ3020: evidence for expression of ETB receptors in human vascular smooth muscle. *J. Cardiovasc. Pharmacol.* **22(S8)**, 22–25.
8. Molenaar, P., O'Reilly, G., Sharkey, A., Kuc, R.E., Harding, D.P., Plumpton, P., et al. (1993) Characterization and localization of endothelin receptor sub-types in the human atrio-ventricular conducting system and myocardium. *Circ. Res.* **72,** 526–538.
9. Davenport, A. P., O'Reilly G., and Kuc, R. E. (1995) Endothelin ETA and ETB mRNA and receptors expressed by smooth muscle in the human vasculature: majority of the ETA sub-type. *Br. J. Pharmacol.* **114,** 1110–1116.
10. Maguire, J. J., Kuc, R. E., and Davenport, A. P. (2000) Orphan-receptor ligand human urotensin ii: receptor localization in human tissues and comparison of vasoconstrictor responses with endothelin-1. *Br. J. Pharmacol.* **131,** 441–446.
11. Maguire, J. J., and Davenport, A. P. (2002) Is urotensin-II the new endothelin? *Br. J. Pharmacol.* **137,** 579–588.
12. Maguire, J. J., Kuc, R. E., Wiley, K. E., Kleinz, M. J., and Davenport, A. P. (2004) Cellular distribution of immunoreactive urotensin-II in human tissues with evidence of increased expression in atherosclerosis and a greater constrictor response of small compared to large coronary arteries. *Peptides* **25,** 1767–1774.
13. Singh, G. Maguire, J. J., Kuc, R. E., Wiley, K. E., Fidock, M., and Davenport, A. P. (2004) Identification and cellular localization of NPW (GPR7) receptors for the novel neuropeptide W-23 by [^{125}I]-NPW-23 radioligand binding and immunocytochemistry. *Brain Res.*, **1017,** 222–226.

14. Katugampola, S. D., Maguire, J. J., Matthewson, S. R., and Davenport, A. P. (2001) [^{125}I]-(Pyr^1)apelin-13 is a novel radioligand for localizing the APJ orphan receptor in human and rat tissues with evidence for a vasoconstrictor role in man. *Br. J. Pharmacol.* **132,** 1255–1260.
15. Kleinz, M. J. and Davenport, A. P. (2004) Immunocytochemical localization of the endogenous vasoactive peptide apelin to human vascular and endocardial endothelial cells. *Regul. Pept.* **118,** 119–125.
16. Katugampola, S. D., Pallikaros, Z., and Davenport, A. P. (2001) [^{125}I-His^9]-ghrelin, a novel radioligand for localizing ghs orphan receptors in human and rat tissue: Up-regulation of receptors with athersclerosis. *Br. J. Pharmacol.* **134,** 143–149.
17. Kuc, R. E., Davies, I. C. and Davenport, A. P. (2003) Motilin receptors in the human cardiovascular system. *Br. J. Pharmacol.* **138,** 165P.
18. Wiley, K. E., and Davenport, A. P. (2003) Detection of CRF2 receptors in human heart using [^{125}]-antisauvagine 30. *Br. J. Pharmacol.* **138,** 16P.
19. Telemaque-Potts, S., Kuc, R. E., Yanagisawa, M., and Davenport, A. P. (2000) Tissue-specific modulation of endothelin receptors in a rat model of hypertension. *J. Cardiovasc. Pharmacol.* **36(S1),** 122–123.
20. Telemaque-Potts, S., Kuc, R. E., Maguire, J. J., Ohlstein, E., Yanagisawa, M., and Davenport, A. P. (2002) Elevated systemic levels of endothelin-1 and blood pressure correlate with blunted constrictor responses and downregulation of endothelin A, but not endothelin B, receptors in an animal model of hypertension. *Clin Sci (Lond).* **103 Suppl 48,** 357S–362S.
21. Davenport, A. P. and Kuc, R. E. (2004) Downregulation of ET_A receptors in ET_B receptor deficient mice. *J. Cardiovasc. Pharmacol.* **43, Suppl 1**, in press.
22. Kuc, R. E., and Davenport A. P. (2000) Endothelin-A-receptors in human aorta and pulmonary arteries are down regulated in patients with cardiovascular disease: An adaptive response to increased levels of ET-1? *J. Cardiovasc. Pharmacol.* **36(S1)**, 377–379.
23. Kutzler, M. A., Molnar, J., Schlafer, D. H., Kuc, R. E., Davenport, A. P. and Nathanielsz, P. W. (2003) Maternal dexamethasone increases endothelin-1 sensitivity and endothelin a receptor expression in ovine foetal placental arteries. *Placenta* **24,** 392–402.
24. Davenport, A. P. (2002) International Union of Pharmacology. XXIX. Update on endothelin receptor nomenclature. *Pharmacol. Rev.* **54,** 219–226.
25. Davenport, A. P. and Russell, F. D. (2001) Endothelin converting enzymes and endothelin receptor localisation in human tissues. *Hdbk. Exp. Pharmacol.* **152,** 209–237.
26. Davenport, A. P., Kuc, R. E., Fitzgerald, F., Maguire, J. J., Berryman, K. and Doherty, A. M. (1994) [^{125}I]-PD15242, a selective radioligand for human ET_A receptors. *Br. J. Pharmacol.* **111,** 4–6.
27. Davenport, A. P., Kuc, R. E., Hoskins, S. L., Karet, F. E. and Fitzgerald, F. (1994) [^{125}I]-PD151242: a selective ligand for endothelin ET_A receptors in human kidney which localises to renal vasculature. *Brit. J. Pharmacol.* **113,** 1303–1310.

28. Peter, M. G. and Davenport, A. P. (1995) Selectivity of [^{125}I]-PD151242 for the human, rat and porcine endothelin ET_A receptors in the heart. *Brit. J. Pharmacol.* **114,** 297–302.
29. Hulme, E. (1992) Receptor-Ligand Interactions. IRL, Oxford, UK.
30. Kenakin, T. (1993) Pharmacologic Analysis of Drug-Receptor Interactions. Raven Press, New York, USA.
31. Kenakin, T. (2004) Principles: receptor theory in pharmacology. *Trends Pharmacol. Sci.* **25,** 186–192.
32. Neubig, R. R., Spedding, M., Kenakin, T., and Christopoulos, A. (2003) International Union of Pharmacology Committee on Receptor Nomenclature and Drug Classification. International Union of Pharmacology Committee on Receptor Nomenclature and Drug Classification. XXXVIII. Update on terms and symbols in quantitative pharmacology. *Pharmacol. Rev11.* **55,** 597–606.
33. Munson, P. J. and Rodbard, D. (1980) Ligand: a versatile computerized approach for characterization of ligand-binding systems. *Anal. Biochem.* **107,** 220–239.
34. McPherson, G. A. (1985) Analysis of radioligand binding experiments. A collection of computer programs for the IBM PC. *J. Pharmacol. Methods* **14,** 213–228.
35. Davenport, A. P. and Hall, M. D. (1988) Comparison between brain paste and polymer standards for quantitative receptor autoradiography. *J. Neurosci. Meth.* **25,** 75–82.

6

Measurement of Radioligand Binding by Scintillation Proximity Assay

Jenny Berry and Molly Price-Jones

1. Introduction

Scintillation proximity assay (SPA) is a homogeneous assay technology *(1)* which is bead-based and removes the need for a filtration step to separate bound from free ligand in a receptor-binding assay. The principle of the technology is shown in **Fig. 1**.

There are two core bead types based on polyvinyltoluene (PVT) and yttrium silicate (YSi) which are both between 2 and 5 μm in diameter. PVT acts as a solid solvent for the scintillant diphenylanthracene, and YSi beads derive their scintillant properties from cerium ions within the crystal lattice. Both bead types, when stimulated, emit light of around 420 nm wavelength, which can be detected by the photomultiplier tubes of a scintillation counter. The surfaces of the beads are covalently coated with a range of different coupling molecules, which enable the receptor to be bound to the bead.

Isotopes which emit β-particles with a short path length in aqueous solution, such as ^{3}H or ^{125}I, are ideally suited for use with the beads, as they will stimulate the bead to which they are bound, but not adjacent beads. Although ^{125}I is a γ emitter, it also gives off two low-energy Auger electrons as part of the decay process, which are detected by the beads.

SPA, therefore, allows the rapid and sensitive assay of a wide range of molecular interactions in a homogeneous system *(2)*. It is routinely used for radioligand-binding assays, particularly in drug screening applications where high throughput is required. Existing filter-binding assays may be readily converted to SPA assays, or assays may be directly developed in SPA format.

From: *Methods in Molecular Biology, vol. 306: Receptor Binding Techniques: Second Edition*
Edited by: A. P. Davenport © Humana Press Inc., Totowa, NJ

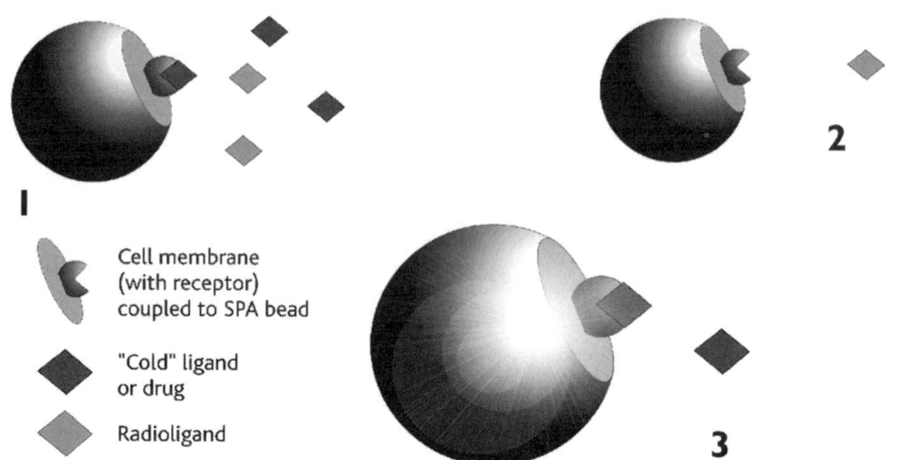

Fig 1. The principle of scintillation proximity assay (SPA). Receptor is bound to the surface of the SPA bead. **(1)** When an unlabeled ligand ("cold") is bound to the receptor, the bead is not stimulated and no light is emitted. **(2)** No ligand is bound; nonbound radioligand does not stimulate the bead. **(3)** When radioligand binds to the receptor, it is in close enough proximity to stimulate the bead and emit light.

This chapter describes the development of SPA radioligand-binding assays detailing the choice of isotope, selection of SPA bead type, optimization of SPA bead and receptor ratio, optimization of assay buffer, selection of assay format, and assay validation including saturation-binding, competition-binding, and association/dissociation-binding studies using SPA (*see* Chapter 5 for analysis of binding data).

2. Materials

1. Radioligand: radioligand labeled with ^{3}H or ^{125}I.
2. Receptor: soluble, membrane preparation, or whole cells may be used.
3. SPA scintillation beads: SPA beads (http://www.amershambiosciences.com) selected from the following range:
   a. SPA scintillation Select-A-bead Kit (Amersham Biosciences, RPNQ0250).
   b. Wheat germ agglutinin (WGA) PVT (Amersham Biosciences, RPNQ0001).
   c. WGA YSi (Amersham Biosciences, RPNQ0011).
   d. Polylysine YSi (Amersham Biosciences, RPNQ0010).
   e. WGA PEI Type A (Amersham Biosciences, RPNQ0003).
   f. WGA PEI Type B (Amersham Biosciences, RPNQ0004).
   g. PEI PVT (Amersham Biosciences, RPNQ0097).
   h. Streptavidin PVT (Amersham Biosciences, RPNQ0007).
   i. Streptavidin YSi (Amersham Biosciences, RPNQ0012).

j. Antirabbit PVT (Amersham Biosciences, RPNQ0016).
k. Antirabbit YSi (Amersham Biosciences, RPN140).
l. Antimouse PVT (Amersham Biosciences, RPNQ0017).
m. Antimouse YSi (Amersham Biosciences, RPN141).
n. Antisheep PVT (Amersham Biosciences, RPNQ0018).
o. Antisheep YSi (Amersham Biosciences, RPN142).
p. Antiguinea Pig PVT (Amersham Biosciences, RPNQ0178).
q. Protein A PVT (Amersham Biosciences, RPNQ0019).
r. Protein A YSi (Amersham Biosciences, RPN143).
s. Glutathione PVT (Amersham Biosciences, RPNQ0030).
t. Glutathione YSi (Amersham Biosciences, RPNQ0034).
u. Copper chelate (His-tag) PVT (Amersham Biosciences, RPNQ0095).
v. Copper chelate (His-tag) YSi (Amersham Biosciences, RPNQ0096).

4. Microplate or tubes suitable for scintillation counter.
5. Protease inhibitors: protease in¡hibitor cocktails (available, for example, from Sigma P-8340 [http://www.sigmaaldrich.com], or Roche Molecular Biochemicals 1 873 580, www.roche-applied-science.com).
6. Harvesting buffer 1: 20 m*M* HEPES, pH 7.4 ± 50 µL/mL protease inhibitor cocktail.
7. Harvesting buffer 2: 50 m*M* Hepes, pH 7.4, containing 145 m*M* NaCl, 2.5 m*M* $CaCl_2$, 1 m*M* $MgCl_2$, 10 m*M* glucose, 0.1% (w/v) BSA, 0.25 mg/mL bacitracin, 0.25 mg/mL Pefabloc SC, 25 µg/mL leupeptin, and 25 µg/mL aprotinin.
8. Tissue homogenization buffer: 50 m*M* Tris-HCl, pH 7.5, containing 2 m*M* ethylenediamine tetraacetic acid (EDTA), 0.25 m*M* sucrose, and 25 m*M* $MgCl_2$.
9. Biotinylation reagents: water-soluble biotin-*N*-hydroxysuccinimide (NHS) esters with variable spacer arm length (PIERCE; http://www.piercenet.com).

3. Methods

3.1. Choice of Isotope

The most suitable isotopes for use in SPA radioligand-binding assays are ^{3}H and ^{125}I, as their decay particles have short path lengths and are therefore only detected if the isotope is bound to the SPA bead. Decay particles from isotopes not bound to the bead will rarely be detected.

The expression level (B_{max}) of the receptor of interest is the primary consideration governing the choice of isotope. In general, B_{max} values <2 pmol/mg membrane protein require the use of [^{125}I]ligands for successful assay development. This is because of the higher specific activity attainable with [^{125}I] labeled ligands. Higher expression levels, e.g., B_{max} >2 pmol/mg, are generally required for successful detection of [^{3}H]ligand binding in SPA assays.

Higher energy isotopes, such as ^{14}C, ^{35}S, and ^{33}P, can be used; however, steps have to be taken to reduce background from unbound isotopes (*see* **Note**

1). These isotopes are not commonly used in receptor-binding SPA assays; however, an SPA assay has been developed for the detection of agonist-stimulated binding of [^{35}S]GTPγS to G-proteins associated with Chinese hamster ovary (CHO) cell membranes expressing receptor protein *(3)*. Chapter 7 describes detection of [^{35}S]GTPγS binding to tissue sections by autoradiography.

3.2. Preparation of Receptor for Use in SPA

3.2.1. Preparation of Cell Membranes

The following method describes the preparation of cell membranes from cells maintained in culture, suitable for use in radioligand-binding SPA assays. Depending on the cell line, addition of protease inhibitors to the harvesting buffer may be required.

1. Wash confluent cell monolayers with cold phosphate-buffered saline (PBS) and remove.
2. Harvest cells by scraping into PBS using a cell scraper.
3. Centrifuge for 10 min at 600*g*.
4. Re-suspend pellet in harvesting buffer 1 and incubate on ice for 10 min to swell cells.
5. Homogenize cells using Dounce homogenizer. Alternatively, cells may be disrupted by brief probe sonication on ice (10 s at maximum power) or other suitable cell-disruption technique.
6. Centrifuge for 10 min at 600*g* and discard the pellet.
7. Centrifuge supernatant for 30 min at 30,000*g*.
8. Re-suspend pellet in storage buffer.
9. Determine protein content of sample.

3.2.2. Preparation of Cell Membranes From Tissue Samples

Cell membranes prepared from tissue samples may be used in SPA radioligand-binding assays. There are no special requirements for preparation of membranes from tissue for use in SPA assays; therefore, it is recommended that the researcher use existing methodologies. *See* **Subheading 3.1.** for information on receptor expression levels required for successful development of SPA radioligand-binding assays.

The following method describes the preparation of porcine lung membranes expressing endothelin receptors suitable for use in SPA binding assay *(4)*.

1. Cut lung tissue into small pieces after removal of trachea and large bronchiole tubes.
2. Homogenize tissue in 1 vol of ice-cold homogenizing buffer using a Waring blender twice for 30 s.
3. Press homogenate through sieve to remove nonhomogenized vessels.
4. Homogenize filtrate using Dounce homogenizer (five passes).

5. Centrifuge homogenate for 10 min at 900*g* (4°C).
6. Carefully remove supernatant and discard. Re-suspend pellet in 1 vol of ice-cold homogenizing buffer.
7. Repeat **steps 5** and **6**.
8. Centrifuge homogenate for 3 hours at 20,000*g* (4°C).
9. Discard supernatant and re-suspend pellet in 2 vol of ice-cold homogenizing buffer and re-homogenize as in **step 4**.
10. Centrifuge homogenate for 30 min at 25,000*g* (4°C).
11. Repeat **steps 9** and **10** twice. Re-suspend final pellet in ice-cold homogenizing buffer.
12. Determine protein content of sample.

3.2.3. Preparation of Whole Cells

The following method describes the preparation of whole cells maintained in culture, suitable for use in radioligand-binding SPA assays. It is essential that protease inhibitors are added to both the harvesting buffer and assay buffer if whole cells are to be used in SPA assays.

1. Wash confluent cell monolayers with cold PBS and remove.
2. Harvest cells by scraping into harvesting buffer 2 using a cell scraper.
3. Centrifuge cells at 600*g* for 10 min and discard the supernatant.
4. Re-suspend the cell pellet in harvesting buffer.
5. Determine protein content of sample.

3.3. Selection of Bead Type

Development of SPA binding assays requires that the receptor be captured onto the surface of the SPA bead. For membrane-bound receptors, this is most commonly achieved using SPA beads coated with WGA, poly-L-lysine (PL), or polyethylene-imine (PEI). Membrane fragments prepared from cells maintained in culture, from tissue, or from whole cells expressing the receptor of interest are directly coupled to the bead via *N*-acetyl glucosamine residues exposed on the surface of the membrane (WGA-bead), or via a charge interaction (PL or PEI bead). Soluble receptors may be biotinylated *(5,6)* and captured using streptavidin-coated SPA beads (*see* **Subheading 3.4.**). Alternatively, glycosylated soluble receptors containing *N*-acetyl glucosamine residues may be directly captured using WGA-coated beads. Soluble receptors may also be captured using secondary antibody or protein A-coated bead in conjunction with a nonneutralizing antibody to the receptor of interest (*see* **Subheading 3.5.**). His-tagged receptors have been captured using copper chelate (His-tag) SPA bead *(7)*, and glutathione-*S*-transferase (GST) fusion protein receptors could be captured using glutathione-coated bead, although to our knowledge not previously performed.

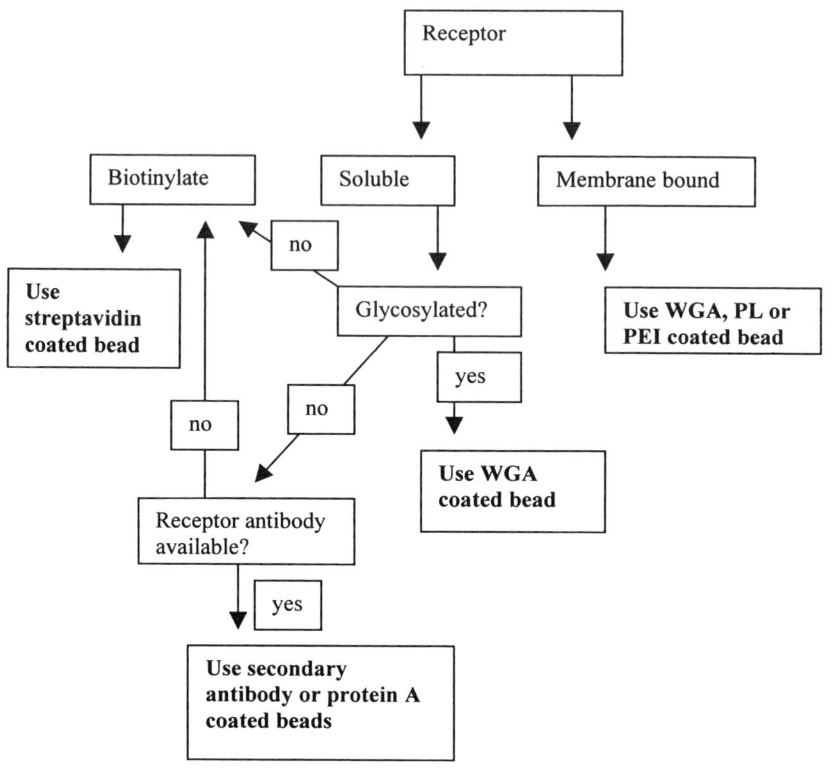

Fig. 2. Flow chart to aid selection of appropriate Scintillation proximity assay bead coating for receptor of interest. WGA, wheat germ agglutinin; PL, poly-L-lysine; PEI, polyethylene-imine.

A flow chart to aid selection of the most appropriate bead coating is shown (**Fig. 2**). The most important factor influencing selection of the optimal core bead type (YSi or PVT) is nonspecific binding (NSB) of the radiolabeled ligand to the bead *(8)*. As a general rule, peptide ligands show a higher level of NSB to YSi SPA beads. Conversely, more hydrophobic nonpeptide ligands tend to show higher NSB to PVT than to YSi beads. The most suitable core bead type can be experimentally determined as described in **Subheading 3.3.1**.

3.3.1. Determination of Nonspecific Binding of Radioligand to SPA Bead

A range of SPA bead types should be used. The "Select-A-Bead" kit is ideal for this purpose if the receptor is membrane-bound. YSi-streptavidin or PVT-streptavidin bead should be used with biotinylated receptors.

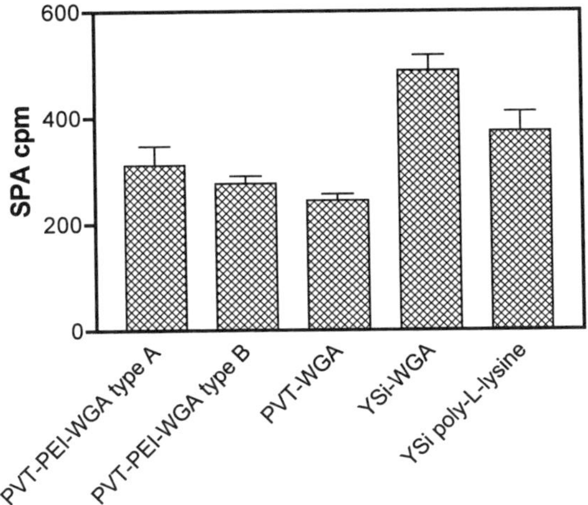

Fig. 3. Binding of [^{3}H]Bradykinin to a range of scintillation proximity assay (SPA) bead types. Values are means ± SD (*n* = 3). In this example, polyvinyltoluene (PVT)-wheat germ agglutinin (WGA) bead was selected for optimization of bead-to- receptor ratio. PEI, polyethylene-imine; YSi, yttrium silicate.

1. Prepare a solution of radiolabeled ligand in assay buffer (*see* **Note 2**) at a concentration equivalent to the approximate K_D of the receptor of interest.
2. Reconstitute SPA bead in distilled water to a concentration of 100 mg/mL.
3. Dilute sufficient bead in distilled water to a concentration of 40 mg/mL.
4. For 100 µL total volume assays, add to suitable microplate or tube: assay buffer (50 µL), radiolabeled ligand (25 µL), SPA bead (25 µL) (*see* **Note 3**).
5. Incubate as required and determine NSB of the radioligand to each bead type using appropriate counting method (*see* **Note 4**).
6. Select bead types with lowest level of bound radioligand for optimization of bead-to-receptor ratio (**Fig. 3**).

3.4. Biotinylation of Soluble Receptors

The following method describes the biotinylation of 50 µg of a lyophilized, purified soluble cytokine receptor using a biotin NHS ester:protein ratio of 5:1. The biotinylated protein is diluted into PBS containing bovine serum albumin (BSA), before dialysis to maximize recovery of the biotinylated receptor protein.

1. Reconstitute 50 µg soluble receptor (carrier-protein free) with 50 µL 50 µ*M* bicarbonate buffer, pH 8.5.
2. Transfer into suitable small container (low-protein binding).

3. Add 10 µL of 0.5 mg/mL solution of water soluble biotin-NHS ester prepared *immediately* before use.
4. Incubate at room temperature for 1 h.
5. Add 10 µL 1 *M* glycine.
6. Incubate at room temperature for a further 30 min.
7. Dilute the biotinylated receptor to 0.5 mL with PBS containing 0.1% BSA (w/v).
8. Dialyze overnight at 2–8°C against PBS.

3.5. Antibody Capture

If the receptor of interest is to be captured using antibody or protein A-coated SPA beads in conjunction with a receptor-specific antibody, optimum concentrations of the receptor, specific antibody, and SPA bead should be determined. It is recommended that this be done in a series of matrix experiments as follows:

1. Using 1 mg/well appropriate antibody- or protein A-bead and a fixed amount of receptor, titrate the specific antibody in the SPA assay. This should be done over a large range of antibody concentrations.
2. From the data generated above, select the range of antibody concentrations giving the largest specific signal (total binding–NSB)
3. Perform matrix experiment titrating both the receptor and antibody using 1 mg SPA bead/well in the assay.
4. Select the antibody and receptor concentrations giving the largest specific signal for further assay optimization.

If required, the SPA bead amount can be further optimized. This would only be necessary if there was a high level of NSB of the radioligand to the bead (reduce bead if possible), or binding was low (increase bead, antibody, and receptor).

3.6. Optimization of SPA Bead-to-Receptor Ratio

Once the most suitable capture method and core bead type has been selected, the optimum ratio of SPA bead to receptor is determined. The method is described for membrane bound receptors captured with WGA- PL- or PEI- coated bead; however, the general method is applicable to soluble receptors, the only difference being concentration of protein used, because of the different capacity of the alternative bead coatings (*see* **Note 5**).

Optimization of the bead-to-receptor ratio is the most important step in the development of SPA binding assays. The aim is to determine the amount of bead and receptor that, added together in the assay well, generate the maximum assay window (in which the assay window is defined as maximum total binding compared with NSB). It is also important to establish that neither the bead nor the receptor are present in excess.

The following method is a recommended starting point for optimization of the bead-to-membrane ratio and is based on a WGA bead capacity of 10 μg membrane/mg bead. Although this appears to be a general rule, it does not apply for every membrane preparation; therefore, following the experiment described, the concentrations of bead and membrane may need to be adjusted, and the experiment repeated, to determine the optimum amounts required.

3.6.1. Preparation of Reagents

1. Reconstitute lyophilized bead in distilled water to give 100 mg/mL, and further dilute bead in water to give 1.0, 0.5, 0.25, and 0.125 mg/25 μL water.
2. Dilute membrane in assay buffer to give 10, 5, 2.5, and 1.25 μg/25 μL assay buffer.
3. Dilute radioligand in assay buffer to give a final assay concentration equivalent to the K_D of the receptor of interest.
4. Dilute unlabeled ligand in assay buffer to appropriate concentration (usually at least 10X K_D) to establish NSB levels.

3.6.2. Experimental Procedure

For 100 μL total volume assays in appropriate microplates or tubes. For a plate map, *see* **Fig 4**.

1. Pipet 25 μL assay buffer into total binding wells/tubes.
2. Pipet 25μL unlabeled ligand into NSB wells/tubes.
3. Pipet 25 μL radioligand into all wells/tubes.
4. Pipet 25μL membrane (10–1.25 μg) into relevant wells/tubes.
5. Pipet 25 μL buffer into "0" membrane wells/tubes, pipet 25 μL bead (1.0– 0.125 mg) into relevant wells/tubes.
6. Incubate and count at regular intervals to establish time course of binding.
7. Plot bound SPA cpm against bead concentration at binding equilibrium as shown (**Fig 5**).
8. Select the optimum bead to membrane ratio from the data.

In the example shown (**Fig 5**) the optimum bead-to-membrane ratio was 0.5 mg bead:5 μg membrane protein. Increasing the bead to 1 mg/well did not increase the specific signal obtained, indicating that 5 μg membrane protein was fully captured by 0.5 mg bead.

3.7. Optimization of Assay Buffer

SPA assays are compatible with typical assay buffers at concentrations used in traditional radioligand-binding assays (*see* **Note 2**).

Because SPA assays are homogeneous, the main aim of assay buffer optimization is to reduce non-specific binding of the radioligand to the bead. The addition of BSA (0.1–0.5% w/v) and/or NaCl (10–100 m*M*) has been shown to

	Bead (mg/well)	0μg membrane		1.25μg membrane		2.5μg membrane		5μg membrane		10μg membrane		
		1	2	3	4	5	6	7	8	9	10	
A	0.125											Total binding wells
B	0.25											
C	0.5											
D	1.0											
E	0.125											Non-specific binding wells
F	0.25											
G	0.5											
H	1.0											

Fig 4. Plate map indicating positioning of wells in a 96-well microplate for optimization of bead and membrane matrix.

reduce NSB in SPA assays. Detergents such as Tween® 20* are also useful in assays utilizing biotinylated soluble receptors captured with streptavidin bead.

3.8. Selection of Assay Format

SPA assays are commonly configured in one of two formats: addition of all reagents at the start of the assay ($T = 0$), or with pre-incubation of bead and receptor. Receptor and bead may be pre-incubated in the assay well prior to addition of radioligand (delayed addition) or precoupled before addition to the assay well/tube. Precoupling the bead and receptor may promote aggregation of the bead due to crosslinking; this applies particularly when precoupling membrane-bound receptor to SPA bead (*see* **Note 6**). The precoupled format is the preferred format for high-throughput screening assays.

3.8.1. Precoupling of Bead and Receptor

1. Determine the optimum bead to receptor ratio as described in **Subheading 3.6.**
2. Mix SPA bead and receptor at the ratio established.
3. Roller mix at 2–8°C for >30 min.
4. Pipet precoupled bead and receptor into assay well (*see* **Note 6**).

3.9. Time Course of Binding and Stability of Signal

SPA radioligand-binding assays are not stopped, therefore it is important to establish both the time course of binding, and the stability of the assay signal at equilibrium.

---

*Tween is a trademark of ICI Americas, Inc.

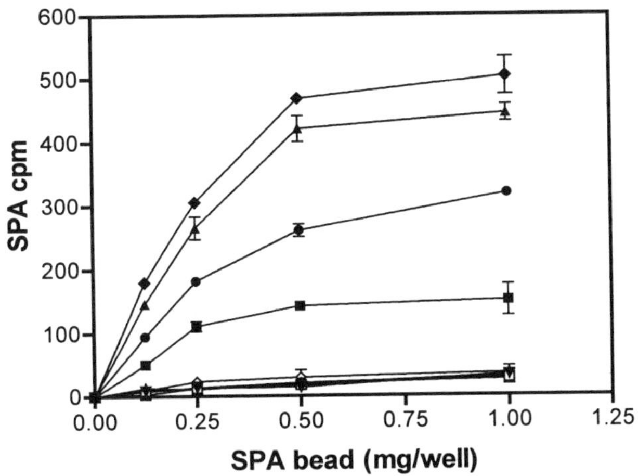

Fig 5. Binding of [N^6-methyl-^{3}H]mesulergine to Chinese hamster ovary-K1 membranes expressing 5HT2_C receptor captured with yttrium silicate (YSi)-wheat germ agglutinin (WGA) bead (data supplied by Euroscreen s.a, Belgium). Optimization of bead-to-membrane ratio. Total binding 0 μg membrane (▼), nonspecific binding (NSB) 0 μg membrane (▽), total binding 1.25 μg membrane (■), NSB 1.25 μg membrane (□), total binding 2.5 μg membrane (●), NSB 2.5 μg membrane (○), total binding 5 μg membrane (▲), NSB 5 μg membrane (◁), total binding 10 μg membrane (◆), NSB 10 μg membrane (◇). Data points are means ± SD (n = 3). In this instance the optimum bead-to-membrane ratio selected from the data was 0.5 mg bead: 5 μg membrane.

This is done simply by setting up a SPA assay using T = 0 format (*see* **Subheading 3.6.**) and counting at regular intervals. Using this method, time to binding equilibrium will include that of receptor binding to the bead. If the time course of ligand binding to the receptor alone is required, the bead and receptor should be pre-incubated or pre-coupled prior to addition of radioligand (*see* **Subheading 3.6**).

Shaking SPA assays during incubation may decrease time to equilibrium; this is particularly apparent with YSi-based SPA bead, as these beads are denser than PVT beads and therefore settle faster.

A decline in radioligand binding on prolonged incubation is occasionally observed (**Fig. 6**). This is often a result of protease activity, and can be resolved by the addition of protease inhibitor cocktails that are commonly used in radioligand-binding assays.

3.10. Assay Validation

Following selection of SPA bead, optimization of bead-to-receptor ratio, assay format, determination of binding equilibrium, and stability of signal, SPA assays can be validated using traditional radioligand-binding methods.

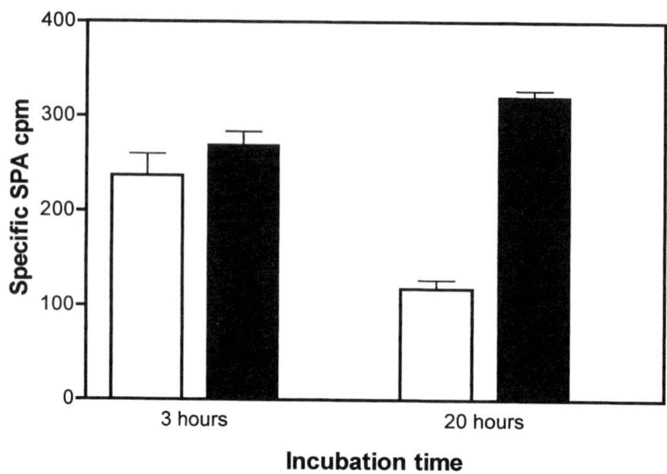

Fig 6. Effect of protease inhibitor addition on [^{3}H]bradykinin binding to Chinese hamster ovary-K1 membranes expressing human B_2 receptor captured with polyvinyltoluene (PVT)-wheat germ agglutinin (WGA) bead. Values are means ± SD (n = 3). No inhibitor (□), inhibitor present (■). Addition of protease inhibitor to the assay buffer prevented loss of radioligand binding following prolonged assay incubation.

3.10.1. Saturation Binding

Both K_D and B_{max} values can be obtained from SPA saturation-binding data. Saturation-binding experiments are carried out in the usual way by increasing the concentration of added radioligand in the SPA assay. The assays are counted at equilibrium to determine total and NSB at each ligand concentration.

The most appropriate way to estimate K_D from SPA binding data is to perform non-linear regression analysis and estimate K_D directly from the binding curve (**Fig. 7**). This avoids any error incurred by determining bound and free ligand.

A B_{max} value in specific bound SPA cpm can also be estimated from the binding curve. Using the counting efficiency appropriate to the core bead type (**Table 1**) the bound SPA cpm, the specific activity of the radioligand, and the amount of receptor protein in the assay well, a B_{max} value in pmol/mg receptor protein can be calculated.

3.10.2. Competition Binding

Competition-binding data can be readily generated (**Fig. 8**). Increasing concentrations of competing ligand are added to the assay, which is incubated to achieve equilibrium, and counted. IC_{50} values can be estimated from the curve in the usual manner ***(9)***.

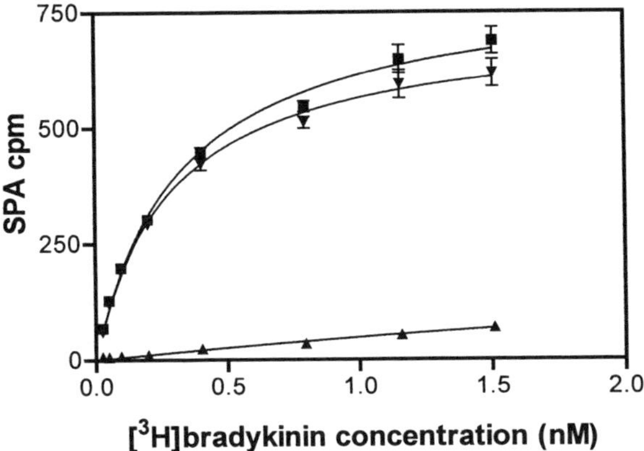

Fig 7. Saturation binding of [^{3}H]bradykinin to Chinese hamster ovary-K1 membranes expressing human B_2 receptor captured with polyvinyltoluene (PVT)-wheat germ agglutinin (WGA) beads. Total binding (■), nonspecific binding (◀), specific binding (▼). Data points are means ± SD (n = 3). Binding curve was fitted using nonlinear regression, and a K_D value of 0.3 n*M* (95% CI, 0.2 to 0.3 n*M*) estimated directly from the curve using GraphPad Prism ***(12)***. A B_{max} value of 729 cpm was estimated directly from the binding curve. Using the counting efficiency of the PVT bead, the specific activity of the [^{3}H]bradykinin, and the amount of membrane protein present in the assay, a B_{max} value of 4 pmol/mg (95% CI, 3.9 to 4.1) was calculated.

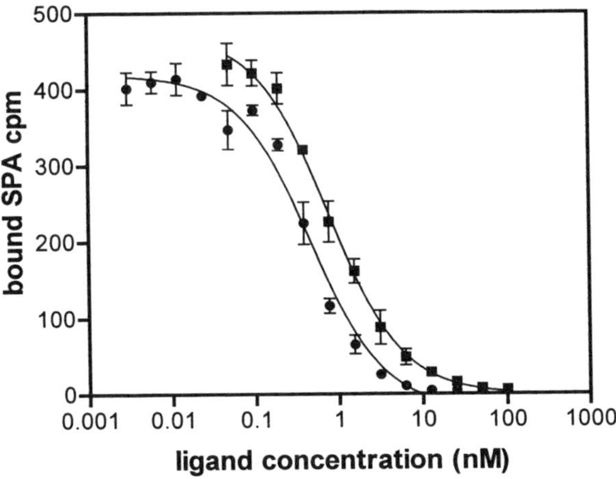

Fig. 8. Binding of [^{3}H]bradykinin to Chinese hamster ovary-K1 membranes expressing human B_2 receptor captured with polyvinyltoluene (PVT)-wheat germ agglutinin (WGA) bead. Competition with Nα-adamantaneacetyl-D-Arg-[Hyp^3,Thi^{5,8},D-Phe^7]-bradykinin (■) and HOE 140 (•). Data points are means ± SD (n = 3). Curves were fitted using nonlinear regression, and IC_{50} values estimated from the curves using GraphPad Prism ***(12)***. Bradykinin IC_{50} 0.78 n*M* (95% CI, 0.39 to 0.58 n*M*) HOE 140 IC_{50} 0.45 n*M* (95% CI, 0.67 to 0.9 n*M*).

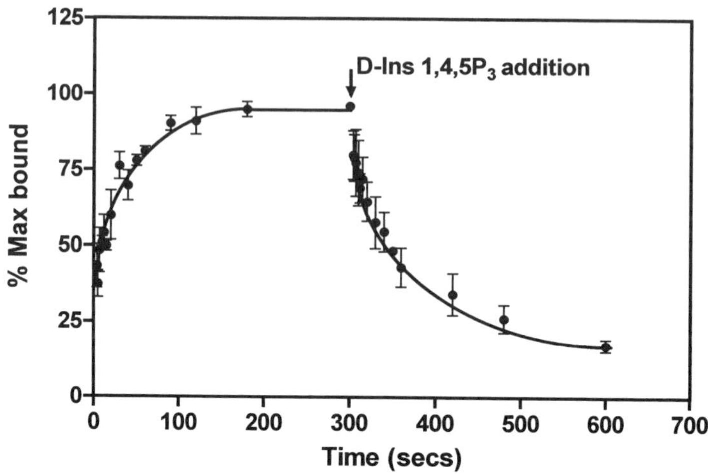

Fig. 9. Association and dissociation of D-[^{3}H]inositol 1,4,5 trisphosphate from solubilized and purified inositol trisphosphate receptors captured with polyvinyltoluene (PVT)-wheat agglutinin (WGA) bead. Precoupled receptor and bead were incubated in the presence of 5 n*M* D-[^{3}H]inositol 1,4,5 trisphosphate for preset times after which the tubes were frozen in liquid N_2 before counting in dry-ice/methanol baths. At equilibrium, excess unlabeled D-Ins 1,4,5P was added to give a final concentration of 1 m*M*. After set incubation times, tubes were frozen and counted as previously described. Data points are means of three separate experiments ± SEM. Data was analyzed using GraphPad Prism ***(12)*** to obtain K_{on} and K_{off}, which were used to calculate a K_D value that was in agreement with the K_D calculated from equilibrium-binding data ***(10)***.

3.10.3. Association/Dissociation Kinetics

Association- and dissociation-binding experiments can also be performed. In fact, SPA is an ideal technology to determine fast on and off rates that may be difficult to obtain using traditional methods ***(10,11)***. Following set-up, SPA assays can be cycle counted at appropriate intervals to achieve binding equilibrium. Ligand dissociation is determined by cycle counting the assay following the addition of a high concentration of competing ligand. From these data, both K_{on} and K_{off} can be calculated in the usual way (**Fig. 9**).

4. Notes

1. Isotopes such as ^{14}C, ^{35}S, and ^{33}P are used in SPA; however, as a result of the longer path length of the decay particles, the bead can be stimulated by isotope not attached to the bead. This results in a high, although constant, background because of this nonproximity effect (NPE).

   NPE can be minimized by pelleting the bead before counting; this is achieved by centrifuging the plate or tubes for 5 min at 500*g*.

Table 1
Comparison of Yttrium Silicate (YSI) and Polyvinyltoluene (PVT) Core Bead Properties and Counting Parameters

	PVT core bead	YSi core bead
Average diameter	5 µm	2.5 µm
Density	1.05 g/cm^3	approx 4 g/cm^3
Settling time	>8 h	30–60 min
Typical counting efficiency compared with liquid scintillation counting	40%	60%
Counting window settings		
Wallac Microbeta™/Trilux	[^{3}H] 5–320 [^{125}I] 5–530	[^{3}H] 5–560 [^{125}I] 5–650
Packard TopCount™	[^{3}H] 1.5–35.00 [^{125}I] 1.5–65.00	[^{3}H] 0.00–50 [^{125}I] 0.00–100
Version 4 software and above	scintillator—liquid/plastic energy range—low mode—normal	scintillator—glass energy range—low mode—high sensitivity
Version 3 software and below	scintillator—liquid energy range—low mode—high	scintillator—solid energy range—low mode—normal
Other instruments	Window wide open	Window wide open

Topcount and Microbeta are trademarks of Perkin Elmer Life Sciences.

2. SPA assays can be performed in any of the assay buffers traditionally used for radioligand binding assays. Lyophilized SPA bead should be reconstituted in distilled water and may be diluted into buffer as required; however, it is preferable to dilute the SPA bead into water for addition to the assay. SPA bead should not be stored once diluted into assay buffer.
   Lyophilized bead can be stored for 12 mo. Once reconstituted, bead should be stored at 2–8°C and consideration should be given to the addition of antibacterial agents, such as sodium azide.
3. It is important to maintain the SPA bead in suspension while pipetting. This is particularly important when pipetting YSi core bead.
4. SPA radioligand-binding assays may be set up in small tubes for counting in conventional scintillation counters, or in microplates suitable for use in microplate scintillation counters.
   When counting SPA assays, the counting window should be set to the appropriate value for the instrument of use (**Table 1**).
   As a result of bead packing, the SPA assay signal will increase as the SPA bead settles to the bottom of the assay well/tube. If a consistent assay to assay signal is required, the assay should be counted once the bead has fully settled (**Table 1**).

5. SPA bead capacities:
   a. PVT and YSi WGA approx 10–20 μg membrane protein/mg bead.
   b. YSi Polylysine, approx 10–20 μg membrane protein/mg bead.
   c. PVT streptavidin, approx 100 pmol biotin/mg bead.
   d. YSi streptavidin, approx 200 pmol biotin/mg bead.
6. Precoupling the SPA bead and membrane protein prior to assay addition may result in aggregation of the bead. This can cause error when pipetting the precoupled bead and membrane. Aggregated bead may be sonicated for approx 30 s using a sonicating water bath before addition to the assay.

References

1. Bosworth, N and Towers, P. (1989) Scintillation Proximity Assay. *Nature* **341,** 167–168.
2. Anon. SPA bibliography, Amersham Biosciences. Available from: http://www1.amershambiosciences.com/aptrix/upp00919.nsf/Content/WD%3ABibliography%2C+S%28190893626-N635%29 [Accessed 27 Feb 2004].
3. Anon. Determination of [^{35}S]GTPgS binding by Scintillation Proximity Assay (SPA). Proximity News 25, Amersham Biosciences. Available from: http://www4.amershambiosciences.com/aptrix/upp00919.nsf/Content/WD%3AProximity+News+%28183960395-R165%29?OpenDocument&hometitle=DrugScr [Accessed 27 Feb 2004].
4. Berry, J. A., Burgess, A. J., and Towers, P. (1991) Scintillation Proximity assay; competitive binding studies with [^{125}I]endothelin-1 in human placenta and porcine lung. *J. Cardiovasc. Pharmacol.* **17(Suppl 7)**, S143–S145.
5. Anon. Biotinylation Techniques and their use in Developing Scintillation Proximity Assays. *Proximity News* **39**, Amersham Biosciences. Available from: http://www4.amershambiosciences.com/aptrix/upp00919.nsf/Content/WD%3AProximity+News+%28148896178-B345%29?OpenDocument&hometitle=DrugScr [Accessed 27 Feb 2004].
6. Anon. SPA Cytokine Receptor Binding assays. *Proximity News* **37,** Amersham Biosciences. Available from: http://www4.amershambiosciences.com/aptrix/upp00919.nsf/Content/WD%3AProximity+News+%28148896913-B345%29?OpenDocument&hometitle=DrugScr [Accessed 27 Feb 2004].
7. Urban, F., Jr., Cavazos, G., Dunbar, J., et al. (2000) The important role of residue F268 in ligand binding by LXRβ. *FEBS Lett.* **484,** 159–163.
8. Anon. Selection of SPA bead type for receptor Scintillation Proximity Assays. *Proximity News* **54,** Amersham Biosciences. Available from: http://www4.amershambiosciences.com/aptrix/upp00919.nsf/Content/WD%3AProximity+News+%28148897012-B345%29?OpenDocument&hometitle=DrugScr [Accessed 27 Feb 2004].
9. Anon. Kinetics of inhibition of receptor binding measured by Scintillation Proximity Assay. *Proximity News* **14,** Amersham Biosciences. Available from: http://www4.amershambiosciences.com/aptrix/upp00919.nsf/Content/WD%3AProximity+ News+%28183972869-R165%29?OpenDocument&hometitle=DrugScr [Accessed 27 Feb 2004].

10. Anon. Characterisation of D-myo-[^{3}H]Inositol 1,4,5 Trisphosphate binding to purified Inositol Trisphosphate Receptors by Scintillation Proximity Assay. *Proximity News* **12,** Amersham Biosciences. Available from: http://www4.amershambiosciences.com/aptrix/upp00919.nsf/Content/WD%3AProximity+News+%28183698040-R165%29?OpenDocument&hometitle=DrugScr [Accessed 27 Feb 2004].
11. Patel, S., Harris, A, O'Beirne, G., et al. (1995) Characterisation of inositol 1,4,5 trisphosphate binding by scintillation proximity assay. *Br. J. of Pharmacol.,* **115,** Proceedings supplement, July 35P.
12. GraphPad Prism version 4.00 for Windows, GraphPad Software, San Diego California USA, www.graphpad.com Copyright (c) 1994–1999 by GraphPad Software. All rights reserved.

7

Autoradiography of Enzymes, Second Messenger Systems, and Ion Channels

David A. Walsh and John Wharton

1. Introduction

Autoradiographic detection of ligand binding to tissue sections has been used to localize, quantify, and characterize a diverse range of sites. Enzymes have been studied via selective inhibitors, ion channels using naturally occurring toxins, and second messenger systems using inositol polyphosphates. Ligand binding complements immunohistochemistry (*see* Chapter 8) and *in situ* hybridization (*see* Chapter 4) by permitting pharmacological characterization and quantification of active sites. Localization, affinity, and specificity of binding sites for ligands (*see* Chapter 5) can be correlated with functional studies performed with the same pharmacological agent. Bioactive ligands are often identified before their targets have been fully characterized, and radiolabeled ligands may become available before molecular and immunological reagents have been developed. A pharmacologically active agent may be synthesized before the endogenous ligand for its binding site has been identified, and autoradiographic methods may help elucidate the site of action of such agents.

Ligand-binding studies are not, however, without their difficulties. Binding depends on the accessibility of functional protein, and may be affected by a wide variety of buffer conditions, the presence of endogenous ligand, or the shedding of binding sites from the tissue surface. Ligand-binding densities, therefore, do not necessarily reflect expression or total concentration of binding protein, but only a snapshot of what is available under the experimental conditions tested. A particular difficulty with many enzymes and inositol

From: *Methods in Molecular Biology, vol. 306: Receptor Binding Techniques: Second Edition*
Edited by: A. P. Davenport © Humana Press Inc., Totowa, NJ

polyphosphate receptors is their often low affinity for available ligands, resulting in elution of specific binding during wash procedures. Furthermore, nonpeptide ligands, such as enzyme inhibitors, often cannot be covalently linked to their binding site, making microscopic localization difficult. Fortunately, maximal binding capacities (B_{max}) are typically several orders of magnitude greater for these ligands than for ligands of peptide receptors, and limitations on the sensitivity of autoradiography can often be overcome by increasing the ligand concentration.

In this chapter, we discuss the autoradiographic detection of enzymes, ion channels, and second messenger systems, illustrated by angiotensin-converting enzyme (ACE), nitric oxide synthase (NOS), vanilloid receptors, guanosine triphosphate (GTP) binding proteins (G proteins), and inositol polyphosphate receptors. Specific protocols described for these sites are intended to illustrate approaches which may be used to develop binding protocols for other systems.

Radiolabeling with ^{125}Iodine can result in high specific activity, but unfortunately not all ligands can be chemically labeled in this way. Furthermore, radioiodination has the potential disadvantage of structurally modifying the ligand, thereby changing its binding characteristics compared with the nonlabeled ligand. Replacement of ^{2}H with ^{3}H retains the structure and chemical characteristics of the nonlabeled ligand. Autoradiographic images of [^{3}H]-labeled ligands typically have higher resolution than [^{125}I]- or [^{35}S]-labeled ligands, but require exposures over periods of weeks or months, rather than days. Ligands containing sulphur groups may be labeled with ^{35}S and require exposures of only hours.

ACE is a dipeptidyl peptidase which converts angiotensin I to the potent vasoconstrictor octapeptide angiotensin II, and which also catalyzes the inactivation of bradykinin and the sensory neuropeptide substance P ***(1)***. Each molecule of endothelial ACE has two zinc-containing catalytic sites, although these may not have identical activities ***(2)***. ACE has been extensively investigated by quantitative autoradiography of radiolabeled inhibitor binding. [^{3}H]-labeled captopril was used in early experiments to study the localization of ACE ***(3)***. Mendelsohn described the use of a radioiodinated tyrosyl derivative of the ACE inhibitor lisinopril, [^{125}I]351A, which has the advantage of higher specific activity ***(4)***. [^{125}I]351A binds to either active site of ACE, and its binding is inhibited by other competitive ACE inhibitors, by ACE substrates such as angiotensin I, bradykinin, and substance P, and by chelators of the active site zinc atom ***(4–6)***. Binding density correlates with enzyme activity, as determined using synthetic substrates in tissue homogenates ***(4,5)***. [^{125}I]351A binds to ACE with nanomolar affinity, but cannot be covalently linked using conventional fixatives. Emulsion-dipping is therefore not possible, but microscopic localization can be achieved by matching film autoradiograms with histochemically stained sections. Antibodies to human and rat ACE are widely available, and the distribution of immunoreactivity closely parallels that of [^{125}I]351A binding ***(7)***. [^{125}I]351A binding permits quantification of the amount of ACE localized to specific tissue structures. Immunohistochemistry is most useful in confirming the cellular localization of ACE. Increased [^{125}I]351A binding may represent upregulated ACE expression, or an increase in the number of cells bearing ACE. A combination of techniques is most appropriate for interpreting such data. [^{125}I]351A binding has been used to demonstrate changes in the amount of tissue-bound ACE during treatment with ACE inhibitors, during pathological processes such atherosclerosis, and during angiogenesis and tissue repair ***(7–10)***.

NO is a free radical with many biological functions, including endothelium-dependent vasodilation and macrophage-dependent cytotoxicity ***(11)***. NO is generated from L-arginine by a family of enzymes known as nitric oxide synthases (NOS). Endothelial and neuronal NOS are constitutively expressed, whereas activated macrophages and some other cell types can express an inducible NOS. Constitutive NOS have high affinities for L-arginine and are mainly calcium-dependent, whereas inducible NOS has lower affinity, higher capacity, and generally lacks calcium dependence. Correspondingly, constitutive NOS generate relatively small amounts of NO in response to specific stimuli, for example, substance P or acetyl choline, whereas inducible NOS can persistently generate large amounts of NO. Binding of the NOS inhibitor N^G-[2,3,4,5-^{3}H]nitro-L-arginine ([^{3}H]-L-NOARG or [^{3}H]L-NNA) provides a quantitative measure of *in situ* NOS activity, but is not entirely specific for any particular isoform ***(12,13)***. The distribution of [^{3}H]L-NNA binding to vascular endothelium and in the central nervous system, in addition to its high affinity, suggest that [^{3}H]L-NNA binds predominantly to constitutive, rather than inducible NOS ***(14,15)***. This view has been supported by comparison with NOS localization by *in situ* hybridization ***(16)***. The recent development of inhibitors with greater selectivity for NOS isoforms has provided valuable tools for the further characterization of NOS-like binding sites ***(14)***. Autoradiographic studies have demonstrated that NOS can be up- or downregulated in vein grafts and during pre-eclampsia ***(15,17)***.

A variety of toxins and pharmacological agents have been identified which bind specifically, and often with very high affinity, to membrane ion channels, through which they mediate their biological effects. Resiniferatoxin binds with high affinity to vanilloid receptors ***(18)***. Resiniferatoxin or capsaicin, the pungent component of red chili pepper, activate vanilloid receptors on fine, unmyelinated sensory nerves, causing them to release neuropeptides including substance P and calcitonin gene-related peptide ***(19,20)***. Sustained stimulation is neurotoxic for sensory nerves. Until the molecular identification of the vanilloid receptor-1 (VR-1), [^{3}H]resiniferatoxin binding was one of the few available methods for studying these sites ***(21,22)***. Even after the development of specific antisera to VR-1, [^{3}H]resiniferatoxin binding can provide data that are complementary to those derived from immunohistochemical techniques. For example, in the dorsal horn of the spinal cord, [^{3}H]resiniferatoxin binding provides data on the average density of receptor, whereas VR-1 immunohistochemistry provides an index of the number of nerves expressing those receptors. [^{3}H]Resiniferatoxin is lipophilic and special protocols have been developed to minimize nonspecific binding. Autoradiography with [^{3}H]resiniferatoxin has been used to demonstrate changes in vanilloid receptor densities in dorsal root ganglia ***(23)***.

The nonhydrolyzable GTP analogs [^{3}H]-5'guanylylimidodiphosphate ([^{3}H]GppNHp) and [^{35}S]-guanosine 5'-*O*-(3-thiotriphosphate) ([^{35}S]GTPγS) have been used in autoradiographic studies of G proteins in tissues **(Fig. 1)** ***(24–26)***. GTP binds with similar affinities to heterogeneous proteins, including the small GTP-binding proteins such as the *ras* gene product, and the α-subunits of heterotrimeric G proteins, which mediate signal transduction by the seven transmembrane family of cell surface receptors ***(27,28)***. Identification of the particular G proteins responsible for altered [^{35}S]GTPγS binding is best performed by using specific antisera on membrane preparations following gel electrophoresis, or by Northern blotting and *in situ* hybridization.

Since the pioneering work of Sim et al. ***(26)***, [^{35}S]GTPγS binding has now been extensively utilized for the localization, characterization, and quantification of receptor-G protein coupling ***(26,29)***. Agonists enhance [^{35}S]GTPγS binding to receptor-coupled G proteins by increasing their affinity for GTP and its analogs, and by decreasing their affinity and thereby encouraging dissociation of guanosine diphosphate (GDP) ***(30)***.

Agonist-enhanced [^{35}S]GTPγS binding has been extensively used for the anatomical localization of G protein-coupled receptors (GPCRs) and regional quantification of receptor-G protein coupling efficacy, and also for the pharmacological characterization of receptors in adjacent tissue structures. Agonist-enhanced [^{35}S]GTPγS binding can be regarded as a functional measure. It compliments data from autoradiography with radiolabeled receptor antagonists or immunohistochemistry, which can localize receptors irrespective of their function. Enhancement of [^{35}S]GTPγS binding can be evaluated for novel agonists whose receptors are not known, or where high-affinity receptor ligands are not available ***(31)***. The technique can discriminate between partial and full agonists, and antagonists. It may also be used to explore potential interactions between different receptor systems ***(32)***. On the other hand, [^{35}S]GTPγS binding may be relatively specific for pertussis toxin-sensitive G proteins (G_0, G_i)

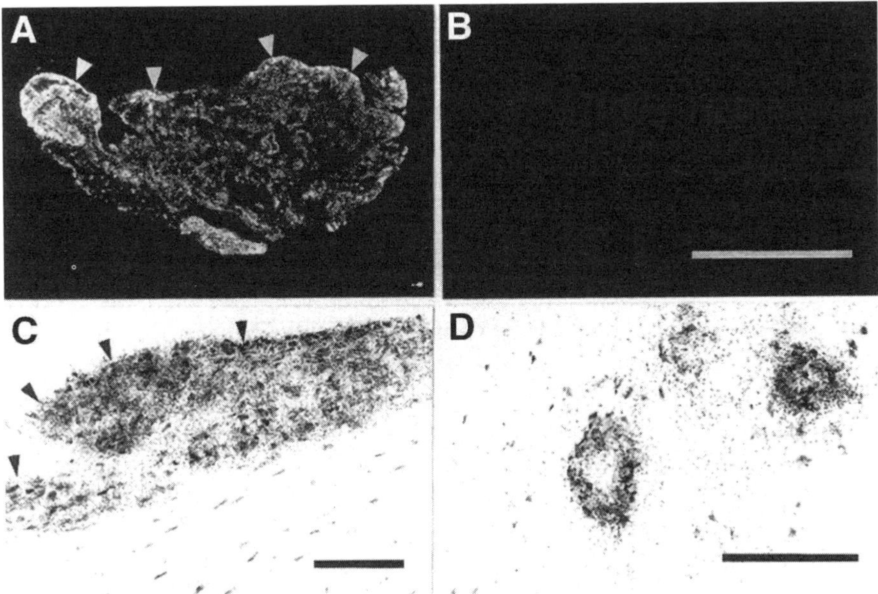

Fig. 1. Localization of $[^{35}S]GTP\gamma S$ binding to human synovium. (**A**) Binding of 0.1 n*M* $[^{35}S]GTP\gamma S$ to the lining region (arrowheads) and blood vessels (arrows), with less dense binding to intervening stroma in synovium from a patient with rheumatoid arthritis. (**B**) Nonspecific binding to a section consecutive to that shown in **A** in the presence of an excess (1 μ*M*) unlabeled GTPγS. (**C**) Dense specific binding indicated by silver grains overlying lining cells in synovium from a patient with osteoarthritis with less dense binding to underlying stroma. Arrowheads indicate synovial surface. (**D**) Specific binding to blood vessels in synovium from a patient with osteoarthritis. The autoradiographic method permits the simultaneous localization and characterization of binding to different structures within a tissue. The density of $[^{35}S]GTP\gamma S$ binding sites was greater on lining cells (B_{max} 47 [95% CI, 22 to 101] fmol mm^{-2}) and blood vessels (B_{max} 39 [95% CI, 29 to 52] fmol mm^{-2}) than on stroma (B_{max} 10 [95% CI, 2 to 43] fmol mm^{-2}, $p < 0.05$), but did not differ significantly among synovia from patients with rheumatoid arthritis ($n = 6$), osteoarthritis ($n = 8$), and chondromalacia patellae ($n = 7$). Note that B_{max} values for $[^{35}S]GTP\gamma S$ are approx 1000-fold greater than for typical G protein-coupled receptors on blood vessels, such as NK_1 receptors for substance P and AT_1 receptors for angiotensin II. These agonists did not significantly affect $[^{35}S]GTP\gamma S$ association or dissociation rates, equilibrium binding, or inhibition by guanosine diphosphate (GDP) in these experiments. **A** and **B**: Reversal prints of film autoradiograms. Bar = 3 mm. **C** and **D**: Emulsion dipped preparations. Bar = 100 μm.

(29). It is not possible, therefore, at present, to investigate all GPCRs with this methodology.

Agonist-enhanced [^{35}S]GTPγS binding has been most extensively used for central nervous system receptors, including opiods, cannabinoids, adenergic, dopaminergic, histaminergic, GABA-ergic, and cholinergic receptors ***(29)***. More recently, the technique has been used to study neuropeptide Y receptor subtypes ***(33,34)***. The applicability of this technique to peripheral tissues has not been fully investigated, although preliminary studies in mice and guinea pigs have demonstrated its feasibility ***(35)***. Agonist-enhanced [^{35}S]GTPγS binding has proved useful for investigating changes in receptor G protein coupling during fetal and adult growth, drug dependence and tolerance, and in the spinal cord during diabetes and nerve injury ***(29,36,37)***.

Other approaches to investigating receptor-G protein coupling in tissue structures include the inhibition of agonist binding by GTP and its analogs, and by specific antisera which block the interaction of G protein and receptor ***(38,39)***.

Inositol polyphosphates are important second messengers, the best characterized being inositol 1,4,5 trisphosphate (IP_3) which mediates cellular actions of agonists of a broad range of GPCRs, including adrenaline, bradykinin, substance P, and angiotensin II ***(40,41)***. IP_3 is generated together with diacyl glycerol by the action of phospholipase C on membrane-derived phosphatidyl choline. IP_3 binds intracellular proteins (IP_3 receptors), stimulating an increase in intracellular calcium concentrations. IP_3 can itself be catalytically converted to other inositol polyphosphates, for example by phosphorylation to inositol 1,3,4,5 tetrakisphosphate (IP_4) which may, in turn, serve other second messenger functions through interaction with specific binding sites. Binding sites for [^{3}H]-labeled inositol polyphosphates can be localized, quantified, and characterized autoradiographically, and at least some of these are likely to be identical to the receptors which mediate their biological actions ***(42,43)***. Inositol polyphosphate binding sites, however, have diverse characteristics, and their biological roles have not all been elucidated. In particular, the synthetic inositol polyphosphate inositol 1,2,6 trisphosphate (α-trinositol), which is believed not to occur in vivo, bound selectively and specifically with high affinity to sites in vascular and nonvascular smooth muscle in both human and rat ***(42)***. These sites shared some characteristics with [^{3}H]IP_4 binding sites—namely, low pH optimum and independence on divalent cations—but displayed different distributions and specificities for related inositol polyphosphates to [^{3}H]IP_4 binding sites in the same tissues. The biological role for these sites remains to be determined, although mediation of the vasodilator effects of α-trinositol has been proposed. The nature of any endogenous ligand for the α-trinositol binding site also remains to be determined. Experience with nonpeptide ligands for peptide receptors indicates that synthetic ligands may show little structural resemblance to their endogenous counterparts. Autoradiography has been used to investigate changes in inositol polyphosphate binding sites following cerebral ischemia, and to help characterize the sites of action of inositol polyphosphate analogs ***(42,44)***.

Autoradiography may therefore, be useful for the localization and quantification of well-characterized sites such as ACE, providing information which is complementary to that obtained by immunohistochemistry and *in situ* hybridization. In addition, it provides a valuable tool for studying ligands with known biological activity whose sites of action are not known either anatomically or molecularly. Pharmacological characterization permits more direct correlation with functional data than is often possible in immunohistochemical and molecular studies.

2. Materials

1. Thermostatically controlled laboratory space suitable for work with radioactive substances at 4°C to 37°C.
2. Tissue-TEK™ O.C.T. mounting compound and cork discs (e.g., from Histological Equipment Ltd.).
3. Isopentane, liquid nitrogen, thermos flask, and a pyrex or metal beaker.
4. Heat sealer and polythene layflat tubing (e.g., from A1 Packagings Ltd., London, UK).
5. Glass microscope slides (e.g., low iron BDH Premium 406/0184/04) and coverslips (e.g., BDH 406/0188/42).
6. Vectabond™ reagent (Vector Laboratories SP-1800, http://www.vectorlabs.com) and acetone.
7. Cryostat, preferably motorized, capable of cutting frozen sections of reproducible thickness, usually 10 or 20 mm (e.g., from Bright Instrument Co. Ltd, http://www.brightinstruments.com).
8. Radiolabeled ligands (e.g., from Amersham Biosciences, http://www.amershambioscinces.com or PerkinElmer Life and Analytical Sciences, Inc. http://las.perkinelmer.com).
9. Reconstitute according to manufacturers recommendations, aliquot at appropriate concentration, and store at –20°C or lower.
10. Unlabeled ligands, aliquotted at appropriate concentrations and stored at –20°C or lower.
11. Buffers and enzyme inhibitors (*see* **Table 1**).
12. Protease-free bovine serum albumin, fraction V powder (e.g., from Sigma A 3294).
13. Metal slide racks (24 capacity) and dishes (400–500 mL capacity).
14. Humidified incubation chambers. Perspex trays (25 × 25 cm) with lids are ideal. Four rods (e.g., plastic pipets) are attached with adhesive inside the base in parallel pairs 4 cm apart to support microscope slides. Moistened absorbent paper is laid between the two pairs of rods.
15. Hair dryer. High air flow and "cold" settings are essential.
16. Darkroom with safelight (e.g., Kodak 6B or Ilford 902/904 filters).

Table 1
Procedures for Specific Ligands

Ligand	[^{125}I]351A	[^{3}H]L-NNA	[^{3}H]resiniferatoxin	[^{35}S]GTPγS	[^{3}H]IP$_4$
Specific activity (TBq mmol^{-1})	74	1.9	3.2	44	0.9
Concentration	0.03 n*M*	100 n*M*	2 n*M*	0.04 n*M*	5 n*M*
Buffer A	10 m*M* PBS pH 7.4	50 m*M* Tris-HCl pH 7.3	10 m*M* HEPES, 2 m*M* $MgCl_2$, 320 m*M* sucrose, 0.75 m*M* $CaCl_2$, 5.8 m*M* NaCl, 5 m*M* KCl, pH 7.4	50 m*M* Tris-HCl, 3 m*M* $MgCl_2$, 0.2 m*M* EGTA, 100 m*M* NaCl pH 7.4	50 m*M* Na acetate 1 m*M* EGTA pH 5.0
Buffer B (A+)	0.2% BSA	10 μ*M* $CaCl_2$	0.1% BSA	2 m*M* GDP 10 mU/mL adenosine deaminase	–
Pre-incubation	2 × 5 min	1 × 15 min	2 × 5 min	2 × 5 min in A then 15 min in B	2 × 10 min
Incubation	3 h at 22 °C	30 min at 20°C	60 min at 37°C	90 min at 22°C	90 min at 4°C
Rinse at 4°C	2 × 5 min	2 × 5 min	2 × 5 min 20 m*M* Tris-HCl, 0.1% BSA, 0.1% a_1 acid glycoprotein	2 × 5 min	2 × 2.5 min
Fixation	–	–	–	30 min Bouin's solution	–
Microscopic localization	IHC	Coverslip	Coverslip	Dip	Coverslip
Exposure; film slide	4 days at 4°C –	6 wk at 4°C 8 wk at 4°C	4 wk at –20°C 8 wk at –20°C	2–3 d at 4°C 1–4 wk at 4°C	3 wk at –20°C 3 mo at –20°C

Abbreviations: BSA, enzyme-free bovine serum albumin; DTT, dithiothreitol; EGTA; ethylene glycol-bis *N*,*N*,*N*′,*N*′-tetraacetic acid; [^{35}S]GTPγS; [^{35}S]-guanosine 5′-*O*-(3-thiotriphosphate); HEPES, *N*-2-hydroxyethylpiperazine-*N*′-2-ethanesulphonic acid; IHC, antibodies available for immunohistochemistry; IP$_4$; inositol 1,3,4,5 tetrakisphosphate; L-NNA, N^{G}-[2,3,4,5-^{3}H]nitro-l-arginine; PBS, phosphate-buffered saline.

17. Autoradiography film (e.g., Kodak BioMax®, MR-1 is suitable for [^{35}S]GTPγS binding and, in combination with BioMax Transcreen LE screens and long exposures, for [^{3}H]-ligands. Uncoated Hyperfilm™-^{3}H can still be purchased from Amersham Biosciences at the time of this writing, but only in limited, special runs).
18. Autoradiography cassettes with side lever closure (e.g., from Genetic Research Instrumentation Ltd., Dunmow, UK, http://www.gri.co.uk/) rather than clip closure designs.
19. Radioactive standards (e.g., Amersham [^{125}I] Microscales™ RPA 523, or [^{14}C]-labeled standards from American Radiolabelled Chemicals, St. Louis, MO, USA, http://www.arc-inc.com)
20. Nuclear emulsion (e.g., Ilford K5 (http://www.ilford.com) diluted 1:1 in water, or Amersham LM-1 RPN 40). Dipping vessel (e.g. Amersham RPN 39). Light tight box (e.g., Raymond Lamb E/107 takes rack E/99).
21. Wire loop (2–3-cm diameter) made from nickel/chrome or platinum wire (approx 0.5 mm thick).
22. Cyanoacrylate adhesive ("Superglue").
23. Developer (e.g., Kodak D19 no. 5027065) and fixer (e.g., Champion Photochemistry, Brentwood, UK., or Amfix™ no. 80213, diluted 1 + 4 with tap water).
24. Histological staining solutions (e.g., hematoxylin and eosin).
25. 70% and absolute ethanol, xylene, or Histoclear and dibutylphthalate polystyrene xylene (DPX) mounting medium.
26. Microscope equipped for transmitted light and dark-field and epi-illumination.
27. Image analysis system.

3. Methods

3.1. Preparation of Sections

1. Unfixed tissues are mounted to cork blocks, frozen in melting isopentane, and stored at –70°C.
2. Glass microscope slides are pretreated to improve section adhesion, for example with Vectabond™ reagent (Vector laboratories, Peterborough, UK) (*see* **Note 1**).
3. Thin (10 mm) sections of unfixed tissue are cut in a cryostat at –20°C to –30°C and thaw-mounted on prepared slides. Sections are air-dried with silica gel desiccant for 1 h at 4°C then used immediately, or stored at –20°C in sealed bags with silica gel.

3.2. Incubations

1. Sections are preincubated in slide racks in baths containing 400 mL preincubation buffer (buffer A, **Table 1**; *see* **Note 2**) at 22°C.
2. Individual slides are removed from the preincubation buffer, gently tapped on absorbent paper, and excess buffer removed by blotting around the edge of the section, then placed horizontally in an incubation chamber.
3. Sections are loaded with a measured volume of incubation buffer containing either radiolabeled ligand alone or together with unlabeled ligand, then incubated for the appropriate time at the stated temperature (**Table 1**; *see* **Note 3**).

4. Incubations are terminated by tapping the slide on absorbent paper to remove ligand and then rinsing twice by immersion in a slide rack in an excess buffer A for the stated times at 4°C (**Table 1**; *see* **Note 4**).
5. Sections are dipped into ice-cold distilled water then immediately dried under a stream of cold air, using a nonheated hair dryer.

3.3. Preparation of Film Autoradiograms

1. Slides are arranged with the sections facing upwards in an autoradiography cassette, together with appropriate radiolabeled standards (*see* **Note 5**).
2. Under safelight conditions, uncoated radiosensitive film is apposed, emulsion-down, to the sections and the cassette is closed. Seal the cassette in a polythene bag containing dry silica gel (*see* **Note 6**).
3. Expose for an appropriate time (**Table 1**) at 4°C or –20°C away from vibrations and other movement (*see* **Note 7**).
4. Bring the cassette to room temperature before removing it from the bag, and open under safelight conditions. Develop in Kodak D19 at 15–20°C for 3 min, stop in tapwater, fix in Amfix (diluted 1 + 4 in tapwater) for 5 min, then rinse in running cold tapwater for 20 min before hanging to dry.

3.4. Quantification of Film Autoradiograms (see Note 8).

1. The autoradiographic film is illuminated from behind using a stabilized light box in a darkened room, and the image captured and converted to a digital image via a video camera. Blank and opaque areas of film are used to correct for variations in light transmission or illumination of the optical system and a shading correction procedure employed.
2. A standard curve is constructed, selecting images of at least six radioactive standards using the cursor.
3. Specific regions of interest on the autoradiographic images of radiolabeled tissue sections are identified and delineated either by using the cursor or by thresholding according to optical density.
4. The integrated grey level values in these regions are transformed using the standard calibration curve derived for each film, thereby giving the amount of ligand bound.

3.5. Preparation of Microautoradiograms (see Notes 9 and 10)

1. Warm an aliquot of diluted Ilford K5 emulsion and the dipping vial to 42°C. Pour emulsion into the vial.
2. Dip a 22 × 64-mm glass coverslip vertically into the emulsion and immediately withdraw, gently scraping one side against the vial to remove excess emulsion. Place vertically on a drying rack, resting against the scraped side, and leave to dry in the dark for 1 h.
3. Put cyanoacrylate adhesive at one end of the unscraped side of the coverslip and appose to the face of the microscope slide bearing the tissue section to which ligand has been bound. The coverslip should be glued to the frosted part of the slide such that the face which is apposed to the tissue section can later be levered

up. Place in an autoradiography cassette, filling space with "blank" slides. Overlay with card and close the cassette so that coverslip and section are apposed under pressure.

4. Place in a sealed plastic bag with silica gel and expose for the required time at –20°C.
5. Warm to room temperature, then open the cassette under safelight conditions. Apply one paper clip across coverslip and slide at the end which is attached by adhesive to prevent the coverslip detaching from the slide. Apply another paper clip across the opposite end of the slide, inserting its tongue under the coverslip and thereby levering the coverslip away from the tissue section.
6. Place in a slide rack and immerse in Kodak D19 developer for 3 min at 20°C, ensuring that developer gains access to the emulsion by vertical agitation. Stop in tapwater, fix in Amfix diluted 1 + 4 in tapwater, rinse for 20 min in running water, then counterstain, for example with hematoxylin and eosin.
7. Dehydrate through graded alcohols (once in 70% ethanol:30% distilled water, then twice in absolute ethanol) then transfer to an organic medium (e.g., twice in xylene or Histoclear). Drop dibutylphthalate polystyrene xylene (DPX) between the coverslip and tissue section, then remove the paper clips. Before the DPX is completely dry, place the slides between paper and cards in an autoradiography cassette and close under pressure in order to minimize the distance between emulsion and section.

4. Notes

1. We have found Vectabond pretreatment effective for most purposes, although some ligands, such as [^{3}H]-α-trinositol, give high nonspecific binding to the surface of pretreated slides and, in such cases, untreated slides are preferred.
2. For agonist-enhanced [^{35}S]GTPγS binding, a second preincubation (15 min in Buffer B at 22°C) is required to saturate G proteins with GDP. A potential limitation of this technique is the often high basal binding of GTPγS in the absence of exogenously applied agonist. Basal [^{35}S]GTPγS binding can be reduced by preincubation with relatively high concentrations of GDP (e.g., 2 m*M*), thereby blocking GTP binding sites except where agonists facilitate GDP–GTP exchange. A proportion of basal [^{35}S]GTPγS binding may result from intrinsic receptor activity and some authors have found that inverse agonists can inhibit basal binding. For example, SR141716A, an inverse agonist at the cannadinoid CB_1 receptor, inhibited basal [^{35}S]GTPγS binding in brain sections ***(45)***. Endogenous agonists within the tissue section may also contribute to the basal [^{35}S]GTPγS binding. In particular, inhibition of adenosine binding by specific adenosine A_1 receptor antagonists, or by degradation through the addition of adenosine deaminase, may reduce basal [^{35}S]GTPγS binding to brain sections ***(46,47)***. Sodium chloride may also inhibit basal [^{35}S]GTPγS binding. The ratio of agonist enhanced to basal [^{35}S]GTPγS binding is further enhanced by encouraging receptor G protein interactions, for example by using ethyleneglycol tetraacetic acid (EGTA)-buffered magnesium.

3. Although 37°C may be considered "physiological," both ligand and binding site may be more stable at lower temperatures, particularly if the tissue contains (unknown) enzymes capable of their degradation. Failure to reach a stable equilibrium binding was observed as an n-shaped association curved with many peptide ligands at 37°C, and with IP_4 at 20°C. Agonists enhance the rate of GTP–GDP exchange rather than increasing the total available GTP binding sites, and therefore agonist-enhancement of $[^{35}S]GTP\gamma S$ binding is observed under nonequilibrium conditions. Evaporation is also more of a problem at higher temperatures. Incubations at 4°C should be performed with precooled sections and buffers . Ligand is loaded in a cold room. Nonspecific binding of $[^{35}S]GTP\gamma S$ is usually undetectable under experimental conditions for autoradiography. Although this can be determined by co-incubation with an excess of non-radiolabeled GTPγS, most investigators find this unnecessary, when studying agonist-enhanced binding.
4. Nonspecific binding, being of lower affinity, dissociates more rapidly than does specific binding during washing. Where specific binding is of very high affinity ($K_d < 0.1$ n*M*), prolonged washes at room temperature may increase the ratio between specific and nonspecific binding by facilitating dissociation of ligand from low affinity, nonspecific sites, while retaining binding to specific sites. With many enzymes and inositol polyphosphate receptors, affinity for radiolabeled ligands may be much lower than this ($K_d > 10$ n*M*). Under these circumstances, important dissociation of ligand from specific binding sites may occur during even short wash periods. Washes should then be performed in ice-cold buffer for short periods (e.g., twice for 1 min) determined empirically to maximize specific binding.
5. Radiolabeled standards should be of the same thickness as experimental sections, particularly when using radioisotopes with highly penetrating emissions. Ten micrometers represents an infinite thickness to emissions from ^{3}H, but not from ^{125}I. $[^{3}H]$-, $[^{125}I]$-, or $[^{14}C]$-labeled polymer standards can be used for $[^{3}H]$-, $[^{125}I]$- or $[^{35}S]$-labeled ligands, respectively.
6. Uncoated autoradiography films are essential for $[^{3}H]$-labeled ligands, whose emissions will be absorbed by protective coatings.
7. Exposure times depend on the radioisotope, specific activity, and the density of binding. $[^{35}S]$-labeled ligands typically require short exposure times (e.g., hours) whereas $[^{3}H]$-labeled ligands typically require weeks or months.
8. The percentage enhancement of $[^{35}S]GTP\gamma S$ binding by agonist gives some indication of receptor G protein coupling efficacy, but should be interpreted with caution. Percentage enhancement will be influenced by, among other things, incubation time and temperature, initial GDP concentration, basal GTPγS binding, receptor number, and receptor-G protein coupling efficacy. Most of these factors, however, may be held constant between regions in a single section. It is possible, by using autoradiography with receptor ligands on consecutive sections, to calculate a catalytic amplification factor by dividing the apparent B_{max} of enhanced $[^{35}S]GTP\gamma S$ binding by the receptor B_{max} ***(48)***.

Quantification of agonist-enhanced GTPγS binding should be described as the increase in binding as a percentage of basal binding. Reported enhancement ranges from near the lower limits of detection (around 30–40%) to sometimes in excess of 150%. The efficacy of enhancement (E_{max}) is the maximum increase in binding. The potency of an agonist can be determined using varying concentrations of agonist as an EC_{50}. Partial agonists give a lower E_{max} value than full agonists at that receptor. Specificity of the agonist-enhanced effect should be demonstrated using specific receptor antagonists.

9. Covalent linkage of radiolabeled ligands to their binding sites on sections, for example, by fixation in paraformaldehyde, may be the most convenient method for microautoradiography. Unfortunately, however, most nonpeptide ligands cannot easily be "fixed" in this way. One approach is to use ligands with reactive moieties which can crosslink to the tissue, as in photoaffinity labelling. Such procedures have been described in autoradiographic studies, although reduced affinity compared with the parent compound has proved a problem in our hands. Use of emulsion-coated coverslips is described here. With these, the emulsion tends not to be as closely apposed to the section as in dipped preparations, and resolution is correspondingly inferior. However, with care and high-activity ligands, localization to structures of approx 50-µm diameter can be achieved. An alternative method, using a wire loop to apply the emulsion to the tissue section, is described in **Note 8**. To our knowledge, "Stripping film," that is, prepared emulsion film that can be floated onto the surface of sections at low temperature, is no longer commercially available.
10. As an alternative to using emulsion-coated coverslips, a layer of emulsion may be applied to unfixed radiolabeled sections using a wire loop. LM-1 emulsion is warmed to 42°C under safelight conditions, then removed from the water bath and allowed to "semi-gel" at room temperature. A wire loop is dipped into the emulsion and applied to the section when the emulsion in the loop appears uniform and stable as viewed under the safelight. If the emulsion is uneven or appears to flow in the loop, it has not gelled sufficiently. Holding the loop above and parallel to the slide, a gentle tap or blow of air may be required to aid the transfer of the gel onto the tissue section.

References

1. Igic, R. and Behnia, R. (2003) Properties and distribution of angiotensin I converting enzyme. *Curr. Pharm. Des.* **9**, 697–706.
2. Wei, L., Clauser, E., Alhenc-Gelas, F., and Corvol, P. (1992) The two homologous domains of human angiotensin I-converting enzyme interact differently with competitive inhibitors. *J. Biol. Chem.* **267**, 13,398–13,405.
3. Strittmatter, S. M. and Snyder, S. H. (1984) Angiotensin-converting enzyme in the male rat reproductive system: autoradiographic visualization with [^{3}H]captopril. *Endocrinology* **115**, 2332–2341.
4. Mendelsohn, F. A. (1984) Localization of angiotensin converting enzyme in rat forebrain and other tissues by in vitro autoradiography using 125i-labelled mk351a. *Clin. Exp. Pharmacol. Physiol.* **11**, 431–435.

5. Correa, F. M. , Guilhaume, S. S., and Saavedra, J. M. (1991) Comparative quantification of rat brain and pituitary angiotensin-converting enzyme with autoradiographic and enzymatic methods. *Brain Res.* **545**, 215–222.
6. Sun, Y., Diaz-Arias, A. A., and Weber, K. T. (1994) Angiotensin-converting enzyme, bradykinin, and angiotensin ii receptor binding in rat skin, tendon, and heart valves: an in vitro, quantitative autoradiographic study. *J. Lab. Clin. Med.* **123**, 372–377.
7. Sun, Y., Cleutjens, J. P., Diaz-Arias, A. A., and Weber, K. T. (1994) Cardiac angiotensin converting enzyme and myocardial fibrosis in the rat. *Cardiovasc. Res.* **28**, 1423–1432.
8. Sun, Y. and Weber, K. T. (1996) Angiotensin-converting enzyme and wound healing in diverse tissues of the rat. *J. Lab. Clin. Med.* **127**, 94–101.
9. Walsh, D. A., Hu, D. E., Wharton, J., Catravas, J. D., Blake, D. R., and Fan, T. P. (1997) Sequential development of angiotensin receptors and angiotensin I converting enzyme during angiogenesis in the rat subcutaneous sponge granuloma. *Br. J. Pharmacol.* **120**, 1302–1311.
10. Zambetis-Bellesis, M., Dusting, G. J., Mendelsohn, F. A., and Richardson, K. (1991) Autoradiographic localization of angiotensin-converting enzyme and angiotensin ii binding sites in early atheroma-like lesions in rabbit arteries. *Clin. Exp. Pharmacol. Physiol.* **18**, 337–340.
11. Moncada, S., Palmer, R. M., and Higgs, E. A. (1991) Nitric oxide: physiology, pathophysiology, and pharmacology. *Pharmacol. Rev.* **43**, 109–142.
12. Michel, A. D., Phul, R. K., Stewart, T. L., and Humphrey, P. P. (1993) Characterization of the binding of [^{3}H]-L-NG-nitro-arginine in rat brain. *Br. J. Pharmacol.* **109**, 287–288.
13. Kidd, E. J., Michel, A. D., and Humphrey, P. P. (1995) Autoradiographic distribution of [^{3}H]L-NG-nitro-arginine binding in rat brain. *Neuropharmacology* **34**, 63–73.
14. Hara, H., Waeber, C., Huang, P. L., Fujii, M., Fishman, M. C., and Moskowitz, M. A. (1996) Brain distribution of nitric oxide synthase in neuronal or endothelial nitric oxide synthase mutant mice using [^{3}H]L-NG-nitro-arginine autoradiography. *Neuroscience* **75**, 881–890.
15. Rutherford, R. A., McCarthy, A., Sullivan, M. H., Elder, M. G., Polak, J. M., and Wharton, J. (1995) Nitric oxide synthase in human placenta and umbilical cord from normal, intrauterine growth-retarded and pre-eclamptic pregnancies. *Br. J. Pharmacol.* **116**, 3099–3109.
16. Burazin, T. C. and Gundlach, A. L. (1995) Localization of no synthase in rat brain by [^{3}H]L-NG-nitro-arginine autoradiography. *Neuroreport* **6**, 1842–1844.
17. Jeremy, J. Y., Dashwood, M. R., Timm, M., et al. (1997) Nitric oxide synthase and adenylyl and guanylyl cyclase activity in porcine interposition vein grafts. *Ann. Thorac. Surg.* **63**, 470–476.
18. Szallasi, A. and Blumberg, P. M. (1990) Specific binding of resiniferatoxin, an ultrapotent capsaicin analog, by dorsal root ganglion membranes. *Brain Res.* **524**, 106–111.

19. Szallasi, A. and Blumberg, P. M. (1999) Vanilloid (capsaicin) receptors and mechanisms. *Pharmacol. Rev.* **51**, 159–212.
20. Gunthorpe, M. J., Benham, C. D., Randall, A., and Davis, J. B. (2002) The diversity in the vanilloid (TRPV) receptor family of ion channels. *Trends Pharmacol. Sci.* **23**, 183–191.
21. Winter, J., Walpole, C. S., Bevan, S., and James, I. F. (1993) Characterization of resiniferatoxin binding sites on sensory neurons: co-regulation of resiniferatoxin binding and capsaicin sensitivity in adult rat dorsal root ganglia. *Neuroscience* **57**, 747–757.
22. Szallasi, A. , Blumberg, P. M., Nilsson, S., Hokfelt, T., and Lundberg, J. M. (1994) Visualization by [^{3}H]Resiniferatoxin autoradiography of capsaicin-sensitive neurons in the rat, pig and man. *Eur. J. Pharmacol.* **264**, 217–221.
23. Szallasi, A. , Nilsson, S., Farkas-Szallasi, T., Blumberg, P. M., Hokfelt, T., and Lundberg, J. M. (1995) Vanilloid (capsaicin) receptors in the rat: distribution in the brain, regional differences in the spinal cord, axonal transport to the periphery, and depletion by systemic vanilloid treatment. *Brain Res.* **703**, 175–183.
24. Gehlert, D. R. and Wamsley, J. K. (1986) In vitro autoradiographic localization of guanine nucleotide binding sites in sections of rat brain labeled with [^{3}H]guanylyl-5'-imidodiphosphate. *Eur. J. Pharmacol.* **129**, 169–174.
25. Aoki, H., Onodera, H., Yamasaki, Y., Yae, T., Jian, Z., and Kogure, K. (1992) The role of GTP binding proteins in ischemic brain damage: autoradiographic and histopathological study. *Brain Res.* **570**, 144–148.
26. Sim, L. J., Selley, D. E., and Childers, S. R. (1995) In vitro autoradiography of receptor-activated g proteins in rat brain by agonist-stimulated guanylyl 5'-[gamma-[^{35}S]thio]-triphosphate binding. *Proc. Natl. Acad. Sci. USA* **92**, 7242–7246.
27. Fields, T. A. and Casey, P. J. (1997) Signalling functions and biochemical properties of pertussis toxin-resistant G-proteins. *Biochem. J.* **321** , 561–571.
28. Denhardt, D. T. (1996) Signal-transducing protein phosphorylation cascades mediated by ras/rho proteins in the mammalian cell: the potential for multiplex signalling. *Biochem. J.* **318** , 729–747.
29. Sovago, J., Dupuis, D. S., Gulyas, B., and Hall, H. (2001) An overview on functional receptor autoradiography using [^{35}S]GTPgS. *Brain Res. Brain Res. Rev.* **38**, 149–164.
30. Brandt, D. R. and Ross, E. M. (1985) GTPase activity of the stimulatory GTP-binding regulatory protein of adenylate cyclase, G_s. Accumulation and turnover of enzyme-nucleotide intermediates. *J. Biol. Chem.* **260**, 266–272.
31. Waeber, C. and Chiu, M. L. (1999) In vitro autoradiographic visualization of guanosine-5'-*O*-(3-[^{35}S]thio)triphosphate binding stimulated by sphingosine 1-phosphate and lysophosphatidic acid. *J. Neurochem.* **73**, 1212–1221.
32. Tanase, D., Martin, W. A., Baghdoyan, H. A., and Lydic, R. (2001) G protein activation in rat ponto-mesencephalic nuclei is enhanced by combined treatment with a mu opioid and an adenosine A_1 receptor agonist. *Sleep* **24**, 52–62.
33. Shaw, J. L., Gackenheimer, S. L., and Gehlert, D. R. (2003) Functional autoradiography of neuropeptide Y Y_1 and Y_2 receptor subtypes in rat brain using agonist stimulated [^{35}S]GTPgS binding. *J. Chem. Neuroanat.* **26**, 179–193.

34. Hong, W. and Werling, L. (2001) Lack of effects by sigma ligands on neuropeptide Y-induced G-protein activation in rat hippocampus and cerebellum. *Brain Res.* **901**, 208–218.
35. Maruo, J., Yoshida, A., Shimohira, I., Matsuno, K., Mita, S., and Ueda, H. (2000) Binding of [^{35}S]GTPgS stimulated by (+)-pentazocine sigma receptor agonist, is abundant in the guinea pig spleen. *Life Sci.* **67**, 599–603.
36. Chen, S. R., Sweigart, K. L., Lakoski, J. M., and Pan, H. L. (2002) Functional mu opioid receptors are reduced in the spinal cord dorsal horn of diabetic rats. *Anesthesiology* **97**, 1602–1608.
37. Bantel, C., Childers, S. R., and Eisenach, J. C. (2002) Role of adenosine receptors in spinal G-protein activation after peripheral nerve injury. *Anesthesiology* **96**, 1443-1449.
38. Walsh, D. A. , Suzuki, T., Knock, G. A., Blake, D. R., Polak, J. M., and Wharton, J. (1994) AT_1 receptor characteristics of angiotensin analogue binding in human synovium. *Br. J. Pharmacol.* **112**, 435–442.
39. Georgoussi, Z., Carr, C., and Milligan, G. (1993) Direct measurements of in situ interactions of rat brain opioid receptors with the guanine nucleotide-binding protein G_0. *Mol. Pharmacol.* **44**, 62–69.
40. Wilcox, R. A., Primrose, W. U., Nahorski, S. R., and Challiss, R. A. (1998) New developments in the molecular pharmacology of the myo-inositol 1,4,5-trisphosphate receptor. *Trends Pharmacol. Sci.* **19**, 467–475.
41. Cullen, P. J. (1998) Bridging the gap in inositol 1,3,4,5-tetrakisphosphate signalling. *Biochim. Biophys. Acta.* **1436**, 35–47.
42. Walsh, D. A. , Mapp, P. I., Polak, J. M., and Blake, D. R. (1995) Autoradiographic localization and characterization of [^{3}H]a-trinositol (1D-*myo*-inositol 1,2,6-trisphosphate) binding sites in human and mammalian tissues. *J. Pharmacol. Exp. Ther.* **273**, 461–469.
43. Worley, P. F., Baraban, J. M., Colvin, J. S., and Snyder, S. H. (1987) Inositol trisphosphate receptor localization in brain: variable stoichiometry with protein kinase C. *Nature* **325**, 159–161.
44. Nagata, E., Tanaka, K., Gomi, S., et al. (1994) Alteration of inositol 1,4,5-trisphosphate receptor after six-hour hemispheric ischemia in the gerbil brain. *Neuroscience* **61**, 983–990.
45. Sim-Selley, L. J., Brunk, L. K., and Selley, D. E. (2001) Inhibitory effects of SR141716A on G-protein activation in rat brain. *Eur J Pharmacol* **414**, 135–143.
46. Laitinen, J. T. (1999) Selective detection of adenosine A_1 receptor-dependent G-protein activity in basal and stimulated conditions of rat brain [^{35}S]guanosine 5'-(g-thio)triphosphate autoradiography. *Neuroscience* **90**, 1265–1279.
47. Moore, R. J. , Xiao, R., Sim-Selley, L. J., and Childers, S. R. (2000) Agonist-stimulated [^{35}S]GTPgS binding in brain modulation by endogenous adenosine. *Neuropharmacology* **39**, 282–289.
48. Sim, L. J., Selley, D. E., Xiao, R., and Childers, S. R. (1996) Differences in G-protein activation by mu- and delta-opioid, and cannabinoid, receptors in rat striatum. *Eur. J. Pharmacol.* **307**, 97–105.

8

Immunocytochemical Localization of Receptors Using Light and Confocal Microscopy With Application to the Phenotypic Characterization of Knock-Out Mice

Anthony P. Davenport and Rhoda E. Kuc

1. Introduction

This chapter describes the immunocytochemical (ICC) localization of receptors in tissue sections and cells growing in culture. The protocol is illustrated using examples of how site-directed polyclonal antisera can be used to distinguish between subtypes of G protein-coupled receptors (GPCRs; endothelin ET_A and ET_B), *(1)* in sections of human tissue or whole-body sections of mice, as well as mapping previously designated orphan receptors (oGPCR) recently paired with their cognate ligand illustrated by visualizing receptors for the peptide ghrelin (*2*). The methods are applicable to other receptor families.

The distribution of receptors within tissue sections can be visualized by radioligand binding combined with film-based macroautoradiography (*see* Chapter 5) or phosphor imaging (*see* Chapter 10), provided a suitable radiolabeled ligand has been developed, which of course is not the case for the remaining orphan GPCRs, where the cognate endogenous ligand remains to be discovered (*3–5*). However, resolution is generally limited to groups of cells unless more technically challenging techniques of micro- (*6*) or electron microscope autoradiography are used (*7*). In contrast, ICC permits the precise identification of cell types expressing a particular receptor when viewed under either a conventional light microscope or by confocal microscopy (*8*). Subtypes of receptors can be identified and distinguished prior to the development of selec-

From: *Methods in Molecular Biology, vol. 306: Receptor Binding Techniques: Second Edition*
Edited by: A. P. Davenport © Humana Press Inc., Totowa, NJ

tive agonists or antagonists. The expression of splice variants of receptors ***(9)*** can be localized and mapped particularly in pathophysiological tissue. Antisera can be used to follow the interaction of receptors with other proteins in the plasma membrane such as receptor activity-modifying proteins (RAMPS) (***10***), and posttranslational modifications to the amino-acid sequence such as glycosylation. The technique also has the additional advantage of avoiding the use of radioactivity (which requires laboratories equipped for the safe handling of radiolabeled ligands), making the technique more widely available.

1.2. Applications

1.2.1. Orphan Receptors

Robas and Fidock (*see* Chapter 2 and **ref. *11***) describe methods used to identify by high-throughput screening the cognate ligand; once this is known it may be possible to radiolabel the ligand in order to identify the native receptor present in tissue. However, for a significant number of orphan receptors predicted to exist by the human genome (about 160 for oGPCRs), these remain unpaired. Until this occurs, these unpaired receptors can still be mapped by site-directed antisera generated to short (typically 15 amino acid sequences) amino acid sequences deduced from the genetic sequence. The anatomical localization of a receptor to a particular type of cell may give a clue to possible function.

1.2.2. Phenotypic Characterization of Receptor Knock-Out Mice

The deletion of genes encoding receptors has emerged as a powerful tool in understanding the role of a specific receptor in physiological processes (***12***). Receptor ICC can be used to analyze the resulting phenotype. Some of these receptor knock-outs have been shown to be similar to mutations in human receptors, such as the endothelin $(ET)_B$ receptor. Homozygote ET_B knock-out mice exhibit a phenotype different from and nonoverlapping with ET_A-deficient animals, being viable at birth and able to survive for up to 8 wk; although they display aganglionic megacolon (which is a characteristic of Hirschsprung's disease) as a result of absence of ganglion neurons together with a pigmentary disorder in their coats (***13***). Mice are usually sufficiently small for the whole body—minus the head and limbs, which can be cut separately if required—to be sectioned using a conventional cryostat (rather than a specialized whole-body sledge microtome as is required for larger species such as rats). The main advantage is that it is possible to collect representative sections to encompass all the major organs without preselection of the tissue to be studied (***8***). Receptor ICC can be used to confirm that the receptor is not expressed in knock-out animals and to study the effects of deleting the gene for a specific receptor on other transmitter systems.

1.3. Sources and Selection of the Primary Antisera

Antisera directed to receptors are commercially available from a number of sources (for example, Phoenixpeptide.com; Abcam.co.uk; Scbt.com). Alternatively, site-directed antisera can be raised against a specific amino acid sequence of a particular receptor (*see* Chapter 1 for databases) and generated either "in-house" or commercially (***14–16***).

1.4. Choice of Fixative

GPCRs are proteins typically comprising about 300–1000 amino acids, and are predominantly anchored within the plasma membrane. For GPCRs the level of fixation required is generally less than for smaller and more labile molecules. There is no ideal method of fixation for ICC, and the best method should be determined empirically. The following protocol describes two methods for fixing fresh frozen cryostat sections: acetone and 4% formaldehyde. It is recommended that both should be tried when characterizing new antisera. Acetone is a simple organic coagulant of cytoplasmic constituents forming a sponge-like reticulum that is easily penetrated by large molecules such as antisera immunoglobulins. Fixation is preferred because although the receptor protein may be distorted, the amino acid sequence is usually not altered. Furthermore, with antisera raised against short antigenic sequences or epitopes that may be buried within the receptor (rather than directed to the C- or N-terminus), acetone can render the target sequence more accessible. A widely used alternative is formaldehyde (4%). This is known to react particularly with N-terminal amino acids and lysine-side chain acids and probably contributes to fixation by this reagent. The exposure time recommended is comparatively short and the degree of crosslinking is likely to be low, but this fixation method may reduce or abolish staining in which antisera are raised to peptide sequences containing one or more lysine residues or to N-termini of receptors.

1.5. Visualizing the Binding of the Primary Antisera by the Peroxidase-Antiperoxidase Method and Dual Fluorescent Labeling

The unlabeled antibody enzyme (peroxidase-antiperoxidase [PAP]) method (**Fig. 1**) is recommended for the visualization of the primary antisera raised against receptors with 3,3'diaminobenzidine tetrahydrochloride (DAB) as the chromogen. The density of receptors is likely to be much less than that of structural proteins. The amplification steps give greater sensitivity over direct staining methods and the production of an insoluble, brown reaction product at the site of the antigen results in a permanent record. The avidin-biotin method (beyond the scope of this chapter) usually gives a similar sensitivity in detecting the primary antisera.

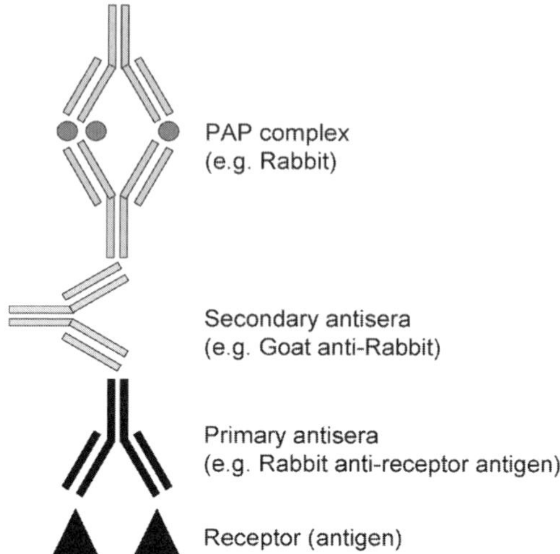

Fig. 1. Schematic diagram illustrating the principle of detecting primary antisera binding to receptors present in tissue sections using the peroxidase-antiperoxidase (PAP) technique leading to the formation of a brown reaction product visible under the light microscope.

Alternatively, the (usually) less sensitive indirect fluorescence method can be used. This is the method of choice for dual labeling of cells. The secondary antisera may be conjugated to a fluorescent dye, such as Alexa Fluor 488, that results in a green fluorescence product under ultraviolet (UV) illumination by light or confocal microscopy. By using another primary antisera raised in a different species (that might be to an endogenous ligand or cell specific marker) visualized with a secondary antisera conjugated to a red fluorophore, such as Alexa Fluor 568, dual labeling in the same section can be used to facilitate identification of cells expressing a particular receptor (**Fig. 2**). Imaging of receptors in living cells is described in detail in Chapter 9.

2. Materials

2.1. Cryostat Tissue Sections

1. Equipment and reagents to snap-freeze tissue, e.g., Dewar containing liquid nitrogen, iso-pentane (2-methyl butane) cooled to the temperature of liquid nitrogen, or aluminium foil cooled and placed on pellets of dry ice (solid CO_2).
2. Single-edged razor blades.
3. Cryostat chucks and cork discs, if used.
4. Mounting medium, OCT compound Gurr® (361603E, VWR International, Poole, Dorset, UK; vwr.com)

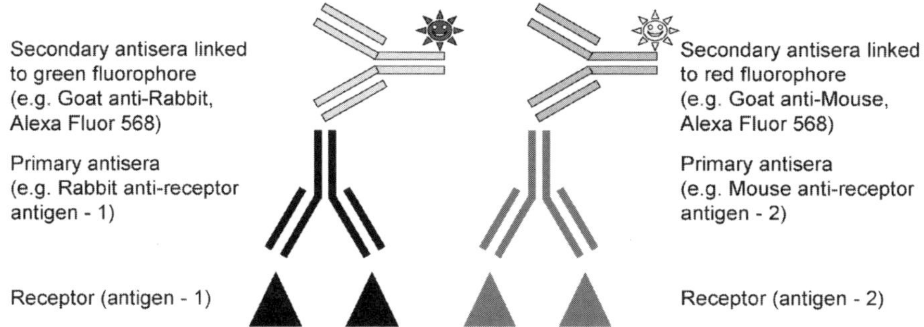

Fig. 2. Schematic diagram illustrating the principal of detecting two primary antisera raised in different species binding to either a receptor or a cell-specific marker in the same tissue section. Secondary antisera linked to fluorophores are used to visualize the primary antisera when imaged using confocal microscopy.

2.2. Fixation

1. Acetone (Analytical grade), ice-cold.
2. 4% Formaldehyde, freshly prepared from an 8% formaldehyde stock solution (*see* **Note 1**).
3. Acetone: methanol, 1:1 mix of analytical grade regents.

2.3. ICC Using the Pexidase-Antiperoxidase Technique

1. Slide racks (metal with handle[s], to hold 24 slides).
2. Slide baths (400–500 mL; note: glass baths with lids should be used for paraformaldehyde, acetone, and xylene, and for other solutions small, plastic lunch/freezer boxes are ideal).
3. Microscope slides coated with poly-L-lysine (e.g., Polysine 406/0178/00; Sigma-Aldrich.com).
4. Sterile gelatin solution for coating coverslips (0.5 g gelatin/200 mL de-ionized water, autoclave and store at 4°C).
5. Phosphate-buffered saline (PBS) (10X stock solution): 400 g NaCl, 10 g KCl, 10 g KH_2PO_4, 57.5 g Na_2HPO_4, dissolved in 5 L of de-ionized water and stored at room temperature). A 1:10 dilution in de-ionized water is made from the stock as required. PBS/T is made by a 1:10 dilution of the stock solution (1 L +9 L de-ionized water) and the addition of 1 mL/L Tween-20 to give a final concentration of 0.1%. PBS/T is stored at 4°C. Other buffers, such as Tris-HCl or a Tris-PBS combination, may be used; however, it is important that they do not contain sodium azide (often found in commercially prepared buffers), as this may inhibit the binding of peroxidase to its substrate and consequently lead to a false-negative result.
6. Hydrophobic pen, such as "Immedge" pen (Vector Laboratories, Peterborough, UK; vectorlabs.com).

7. Incubation trays, e.g., 24 × 24-cm NUNC bioassay dishes (Gibco BRL), modified by the addition of four perspex rods fixed with adhesive to the bases in two pairs, to give support to two rows of six to seven slides per tray.
8. Site-directed primary antisera to target receptor. The protocol is illustrated with receptor antisera raised in rabbit but can be adapted for antisera raised in other species such as rat, sheep, goat, and so on.
9. Secondary antisera illustrated with goat antisera raised against rabbit immunoglobulin (Ig) (FC segment of IgG). Secondary antisera to a range of species are available for a number of suppliers (e.g., DAKO Ltd, Ely, Cambs, UK; dakocytomation.co.uk). This company supplies secondary antisera pre-absorbed against human antigens to reduce nonspecific staining.
10. PAP raised, as in this example, in rabbit (because the primary antisera was raised in this species).
11. Chromogenic substrate (DAB; D-5637, Sigma-Aldrich), made up in hydrogen peroxide-supplemented 0.05 *M* Tris-HCl buffer, pH 7.6 (*see* **Note 2**).
12. Reagents for dehydrating and clearing tissue sections: industrial methylated spirits (IMS), absolute ethanol, and xylene (all analytical grade, from a general laboratory supplier). A series of alcohol baths are prepared for dehydrating sections: 30%, 70%, and 100% IMS (balance de-ionized water); 100% ethanol (×2); and acid alcohol (100% ethanol + 1 mL concentrated HCl). With the exception of 100% ethanol, these may be stored and re-used several times.
13. Reagents for counterstaining (if required). Hematoxylin solution, e.g., Harris' modified hematoxyilin solution (HHS-16, Sigma-Aldrich) and Scott's tap water (3.5 g sodium bicarbonate, 20 g magnesium sulfate added to 1 L of de-ionized water).
14. Mounting medium for coverslips: xylene-based permanent mountant (e.g., DePeX-Gurr, 361254D, VWR International, Poole, Dorset, UK; vwr.com).
15. Light microscope.

2.4. Dual Labeling Using the Indirect Fluorescent Technique and Confocal Microscopy

1. Equipment and reagents to prepare cryostat tissue sections and fixation using acetone or 4% formaldehyde.
2. Site-directed primary antisera to target receptor. The protocol is illustrated with receptor antisera raised in rabbit, with the second primary antisera raised in mouse to a cell-specific marker.
3. PBS/T containing 1% nonimmunized swine serum.
4. Blocking sera (5% nonimmunized swine serum in PBS).
5. Secondary antisera: goat antirabbit conjugated to a green fluorescent dye (Alexa Fluor 488, Molecular Probes; www.probes.com) and goat antimouse conjugate to red florescent dye (Alexa Fluor 568).
6. Microscope cover slips suitable for confocal microscopy (No. 1 thickness)
7. Vectashield antifading mounting medium (Vector Laboratories, Peterborough, UK; vectorlabs.com).
8. Confocal laser scanning microscope (e.g., Leica microscope).

3. Methods

3.1. Testing and Validating Antisera

1. Specificity: primary antisera (raised in-house or from commercial supplier) should be characterized for specificity for its target antigen by enzyme-linked immunosorbent assay (ELISA) (*see* **Note 3**).
2. To enhance the specificity of the antisera for the antigen, it is advisable, where possible, to affinity-purify the antisera to remove nonspecific immunoglobulins and any interfering peptides (e.g., using the SulfoLink Kit from Perbio Science UK Ltd., Cheshire, UK; perbio.com) this has usually already been carried out for commercial antisera. Once the specificity of the antisera is known, positive and negative tissue controls can be included routinely to monitor the assay system. For a novel peptide, this requires the processing of a range of tissues to first determine the distribution of the antigen and thus identify suitable negative and positive tissue controls.
3. Positive controls can include testing antisera using tissue or cells known to express the target receptor from previous studies, or where molecular studies have shown the presence of high levels of mRNA encoding the receptor.
4. Omit the primary antisera in adjacent sections and carry through all staining procedures. Little or no staining should be detected. Any intense staining will be an artefact and could be a result of endogenous peroxidase.
5. Pre-absorb the primary antisera with an excess of the immunizing peptide, and carry through all staining procedures. This should result in the attenuation or complete loss of staining in a tissue, although it is rare to completely abolish staining. This procedure may have been done for antisera from commercial sources, or companies may provide the antigenic peptide.
6. Substitute a nonimmune serum (e.g., normal rabbit serum when rabbits have been used to produce the primary antisera). Little or no staining should be observed. Any intense staining suggests nonimmunological attachment of γ-globulins to the tissue. Ideally, this would be a pre-immune serum sample taken from the same animal prior to challenge with the antigen (with antiserum from a commercial source, this may again not be possible).
7. When determining the distribution of the peptide antigens in a range of human and animal tissues, comparisons with other known cell markers—e.g., von Willebrand factor, smooth muscle α-actin, glial fibrillary acidic protein, and macrophage staining—can be used as positive controls for the assay procedure and also to confirm localization of the antigen to specific cell types. Once the distribution of the antigen is determined, known positive tissues should be included in the protocol when investigating "new" tissues.
8. In ICC, the concentration of antigens and antibodies is crucial to the achievement of a reliable result; too high a concentration of antisera can result in a false-negative (a similar phenomenon to that seen in agglutination testing), or in the more easily predicted overstaining and increase in nonspecific background staining. The ideal ICC protocol is one that achieves a balance, giving the highest

possible specific staining with the least background staining. It is advisable, when using a new antisera or a previously characterized antisera in a new tissue, to use a range of dilutions not only for the primary antisera but also for the secondary or "link" antisera as well as the peroxidase conjugate. A titration checkerboard is a useful example of how such dilutions might be organized.

9. Finally, to demonstrate the localization of the antigen within a tissue section, it is possible to counterstain the sections. The DAB product is insoluble in alcohol and organic solvents; it is therefore possible to use a hematoxylin stain prior to the alcohol dehydration and xylene steps and the mounting of the slides. Development of the counterstain should be monitored to avoid overstaining, which may result in obliteration of the antigen-specific staining.

3.2. Tissue Preparation

3.2.1. Preparation of Tissues and Cryostat Sections

1. Tissues for ICC should be as fresh as possible and processed within minutes of removal. Tissue should be cut into blocks of 1–2 cm^3 using a single-edged razor blade, frozen over dry ice, isopentane, or liquid nitrogen and wrapped loosely in foil, labeled, and stored at –80°C until required (*see* **Note 4**). Alternatively, the tissue may be frozen directly onto cryostat chucks, or cork discs, by embedding in OCT mounting medium either over dry ice or on the cold shelf of the cryostat. The mounted tissue may be sectioned immediately or stored, wrapped in foil, at –80°C (*see* **Note 5**).
2. Allow tissue, whether previously stored at –80°C or freshly mounted, to equilibrate to cutting temperature, i.e., –20 to –30°C depending on tissue type (*see* **Note 6**). For ICC, cut cryostat sections at 30 μm and thaw-mount onto microscope slides pretreated with poly-L-lysine to aid adhesion of the tissue section (*see* **Note 7**). Allow cut sections to air-dry before either commencing ICC or storing slides in sealed boxes at –80°C until required.

3.2.2. Preparation of Cultured Cells for ICC (e.g., HUVECs)

1. Cultured human umbilical vein endothelial cells (HUVECs), at primary passage, are seeded into 12-well culture dishes containing gelatin-coated coverslips and incubated for up to 7 d until almost confluent (*see* **Note 8**).
2. Remove culture plates from the incubator and wash (twice for 5 min) with PBS to remove incubation media and immediately process for ICC (*see* **Note 9**). If cells cannot be processed immediately, fix with a 1:1 mixture of acetone and methanol and freeze at –80°C until required.

3.2.3. Preparation of Whole-Body Sections From Knock-Out Mice

1. Euthanize knock-out and control mice. Remove skin, then limbs and head to assist with the cryostat sectioning. Wrap body loosely in aluminium foil before freezing in a –80°C freezer to avoid cracking. Store at this temperature until required.
2. Mount whole body onto large cryostat chucks using OCT mounting medium, orientated to give the desired transverse, coronal, or longitudinal sections. Cut

cryostat sections at 30 µm and thaw-mount onto microscope slides pretreated with poly-L-lysine to aid adhesion of the tissue section (*see* **Note 7**).

3.3. Immunocytochemistry

3.3.1. Tissue Sections

1. Transfer air-dried sections (dried for between 3 and 24 h) into slide racks.
2. Acetone fixation: fix sections by immersion in ice-cold acetone for 10 min, then air dry.
   or
   Formaldehyde fixation: fix by immersion in 400-mL baths of freshly prepared 4% paraformaldehyde in 0.1 *M* PBS, for 30 min at 4°C (*see* **Note 10**). Wash slides by immersing, in racks, into 400-mL baths of PBS (three times for 5 min), then remove excess buffer by carefully wiping around each section.
3. For both methods of fixation, encircle the sections with a hydrophobic pen and label each slide using a pencil (*see* **Note 11**). Place the slides horizontally into incubation trays humidified by the addition of PBS to the bottom of the trays (*see* **Note 12**).
4. Block nonspecific staining by covering sections with 200–500 µL of 10% "normal" swine serum in PBS (*see* **Note 13**). Incubate sections, in the trays with lids on to maintain humidity, for up to 2 h at room temperature.
5. Tip off the blocking reagent and gently tap the slides on the base of the incubation tray. Wipe any excess carefully from around the sections. Then add a volume (typically 100–300 µL) of the primary antisera against the antigen under investigation (also positive and negative controls), diluted appropriately in PBS/TSS (i.e., PBS, 0.1% Tween–20, + 1% swine serum, 3.3 mg/mL BSA), onto each section, and incubate the slides at 4°C overnight (*see* **Note 14**).
6. Return the slides to the slide racks and wash, as in **step 2**. Return slides to the trays and incubate with the appropriately diluted secondary antisera (swine antirabbit at 1: 200 in PBS/TSS) for 1 h at room temperature (*see* **Note 15**).
7. Wash slides by immersing, in racks, into 400-mL baths of PBS/T (three times for 5 min), remove excess buffer by carefully wiping around each section, and incubate for a further 1 h at room temperature with the appropriately diluted (1:400 in PBS/TSS) rabbit PAP reagent (*see* **Note 16**).
8. Wash slides again as in **step 6**, then incubate for up to 4 min with a freshly prepared solution of the chromogenic substrate DAB. Monitor slides for development of the brown reaction product. Stop the reaction by gently flooding the sections with de-ionized water from a wash bottle, or pipet and transfer them to slide racks in a bath of de-ionized water. At this point, the slides may be counterstained with Hematoxylin to aid in the interpretation of antigen distribution (*see* **Note 2**).
9. Transfer the slide racks through the series of alcohol baths (1–2 min each) to dehydrate the sections, then into a bath of xylene for at least 1 h to clear. Mount the slides using DePeX and coverslips, allow them to dry, and remove excess mountant before viewing under a light microscope (**Figs. 3** and **4**).

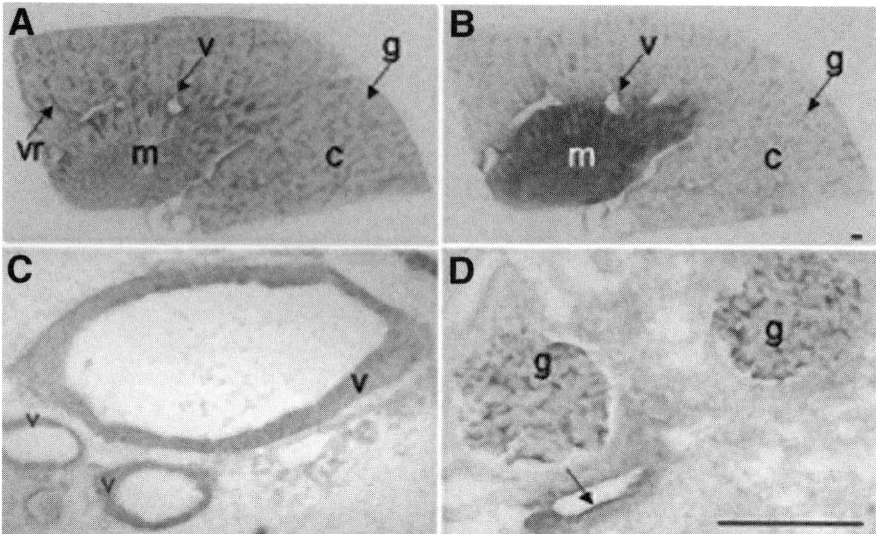

Fig. 3. Brightfield photomicrograph illustrating the cellular localization of receptor subtypes using immunocytochemistry. (**A**) Localization of ET_A receptor IR in human kidney to smooth muscle cells of arcuate arteries and adjacent veins and (**C**) at higher magnification. ET_B receptor immunostaining to endothelial cells lining blood vessels and the vasa recta (**B**). At higher magnification, staining can be detected within endothelial cells of glomeruli (c, cortex; g, glomerulus; m, medulla; v, vessel; vr, vasa recta; [arrows indicate endothelial cells]. Scale bar = 200 µm.

3.3.2. Cultured Cells

1. To fix cells (freshly prepared as in **Subheading 3.2.2.** or, if frozen cells are used, allow to thaw) on coverslips, carefully add 2 mL of acetone:methanol (1:1 mix) to each well, and incubate for 5 min. Aspirate to remove the fixative and allow cells to dry.
2. Carry out the ICC procedure as described for tissue sections (*see* **Subheading 3.3.1.**) with the following modifications: the incubation volume used is 1 mL / well. Wash cells by aspirating the incubation solutions, and replace 2–3 times with PBS/TSS (*see* **Note 9**).
3. Following incubation with DAB, aspirate the wells and fill with de-ionized water to stop the reaction. Remove each coverslip individually from the wells and dip through an alcohol series to dehydrate. Mount each coverslip, cell side down, onto a microscope slide with DePeX and allow to dry before viewing under a light microscope (*see* **Note 17**).

3.3.3. Immunocytochemistry With Immunofluorescence Visualization

1. Process tissue sections or cells as described in **Subheading 3.3.1.** or **3.3.2.,** using either acetone or 4% formaldehyde fixation.

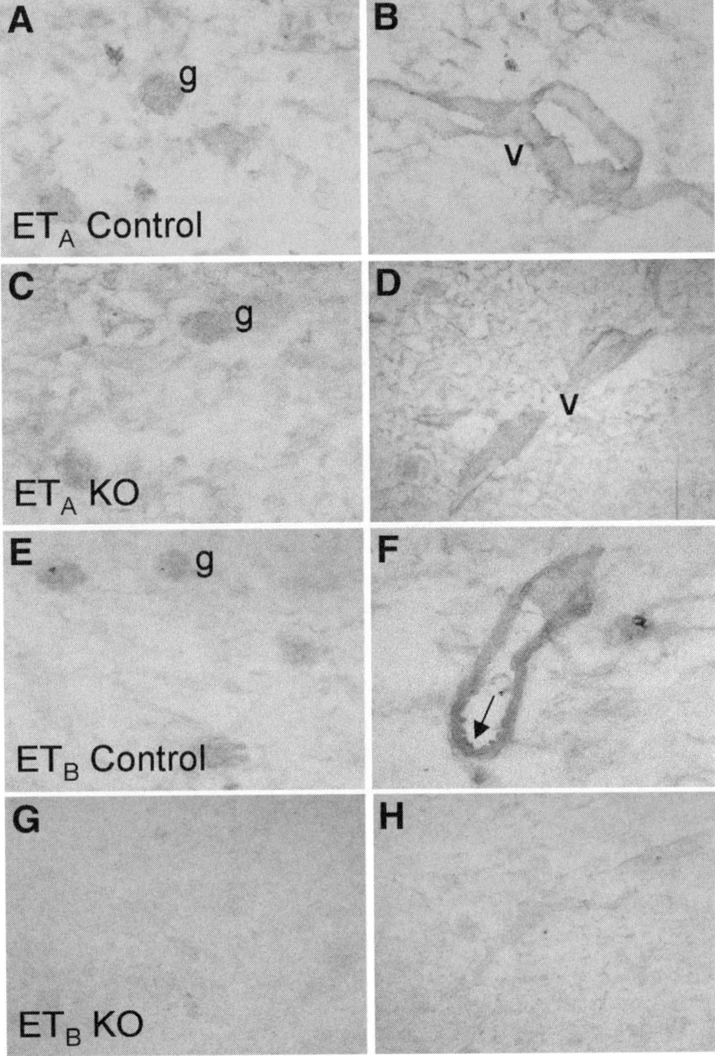

Fig. 4. An example of the phenotypic characterization of receptor knock-out mice, comparing endothelin ET_A and ET_B IR in mouse kidney. A similar distribution of ET_A IR was visualized, as expected, to smooth muscle cells within the renal vasculature from control (+/+) mice (**A**,**B**) and homozygous (–/–) mice in which the ET_B gene had been knocked out. ET_B IR localized to endothelial cells of the glomeruli and as a single layer lining blood vessels (arrows) in controls (**E**,**F**), but staining was not detected (**G**,**H**), as expected, in mice with the deleted ET_B gene (g, glomerulus; v, vessel).

2. Incubate sections with the two primary antisera in PBS containing 1% swine serum. The optimum concentration for each antisera is determined separately by serial dilution. Sections are incubated with the two antisera in PBS containing 1% swine serum for 24 h at 4°C.
3. Wash three times for 5 min in PBS containing 1% swine serum.
4. Secondary antisera: goat antirabbit conjugated to a green fluorescent dye (Alexa Fluor 488) and goat antimouse conjugate to red florescent dye (Alexa Fluor 568), dilute 1:100 in PBS (*see* **Note 18**). Leave for 1 h at room temperature.
5. Wash as described above and mount with Vectashield mounting medium before examining under confocal microscope. Sections can also be examined using a confocal microscope or conventional epifluorescent fluorescent microscope equipped with incident-light fluorescent modules with the following typical specifications. For Alexa Fluor 488 and other dyes with spectra similar to fluorescein isothiocyanate (green): excitation filter band pass 450–495; dichroic mirror 510 nm; selective barrier filter, band pass 520–560. For Alexa Fluor 568 and other dyes with spectra similar to rhodamine isothiocyanate (red): excitation filter band pass 546/10 nm; dichroic mirror 580 nm; long pass filter 590 (*see* **Fig. 5, Note 19**).

4. Notes

1. Prepare an 8% w/v solution by depolymerizing 40 g paraformaldehyde in 500 mL of de-ionized water by heating on a hotplate stirrer in a fume hood to 80°C. Clear solution by adding 1 mL of glacial acetic acid, allow to cool, then filter and store at 4°C. A 400 mL, freshly prepared working solution is made by adding 200 mL of 8% formaldehyde to 200 mL of 2XPBS, i.e., 1:5 dilution of the 10X PBS stock.
2. DAB D-5637, Sigma-Aldrich, Poole, is carcinogenic. Care should be taken to avoid breathing in the powder when handling it. To avoid weighing out the powder, in a fume cupboard add sufficient de-ionized water to dissolve the entire contents of the supplied vial to a concentration of 24 mg/mL. The DAB solution may then be stored in 2.5-mL aliquots at –20°C. To prepare the final incubation solution, add a thawed 2.5-mL aliquot to 100 mL of 0.05 *M* Tris–HCl buffer, pH 7.6, and immediately prior to use add 1 mL 0.3% hydrogen peroxide (this should give a final DAB concentration of 0.6 mg/mL; this solution will remain stable for 1–2 h). The colored product is insoluble in alcohol and organic solvents, thus allowing for counterstaining with the alcohol-based stain Hematoxylin (following wash in water bath incubate for 5 min, wash in tap water for 30 s, followed by Scott's tap water for 1 min. Wash again before continuing with protocol.), dehydration through a series of baths of increasing alcohol concentration, and permanent mounting in xylene-based mountants such as DePeX.
3. ELISA protocol: 96-well plates (NUNC maxisorp) are coated with 100 μL of the immunizing peptide of interest at a concentration of 1 μg/mL in PBS and incubated overnight in a humid box at 4°C. Following washing with PBS/T, nonspecific absorption is blocked by incubating each well with 400 μL of 3% BSA in

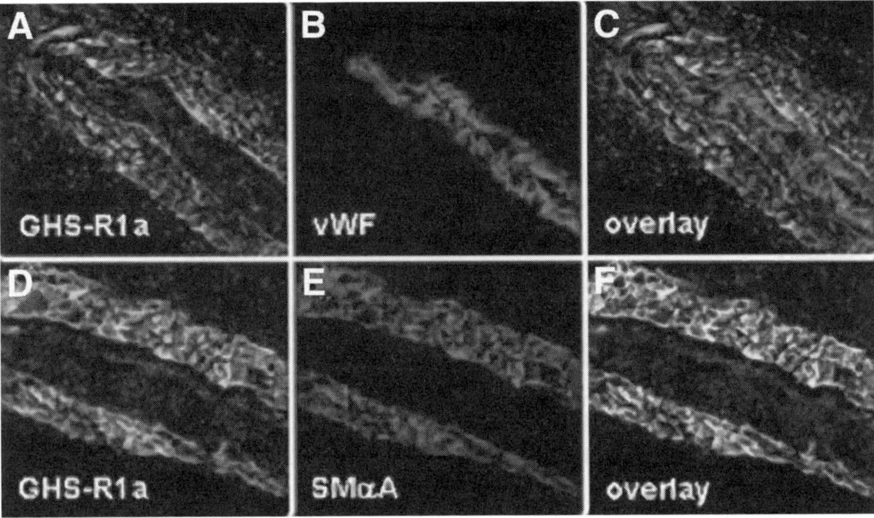

Fig. 5. Confocal photomicrographs illustrating the co-localization of antisera to a receptor (for the peptide ghrelin) and cell-specific markers (for endothelial or smooth muscle cells). Ghrelin receptor IR can be visualized in cross-sections of a human intramyocardial coronary artery (**A** and **D**, detected by green fluorescent secondary antisera). In **B**, a cell-specific marker (von Willebrand Factor) has been used to visualize, in the same section, the single layer of endothelial cells (detected by red fluorescent secondary antisera). In **E**, smooth muscle cells have been visualized by antisera to α-actin (detected by red fluorescent secondary antisera). The digitally overlaid images demonstrate co-localization of ghrelin receptor IR with endothelial cells (**C**) and the smooth muscle layer (**F**).

PBS for 2 h at room temperature. After a further wash, 100 μL of the antisera being tested is added to replicate wells over a range of threefold dilutions from 1:100 (e.g., 1:100, 300, 900, 2700, 8100, 24300, 72900, 218700), and the plates are again incubated overnight as above. The plates are washed and each well incubated with (for antisera raised in rabbits) 100 μL of HRP-conjugated swine anti-rabbit IgG for 2 h at room temperature prior to a final wash and addition of 100 μL per well of a TMB substrate, which produces a blue-colored reaction product. By the addition of 100 μL 1 *M* H_2SO_4, this is converted to a yellow product and the optical density (OD) measured in a plate reader at 450 nm. Both related and unrelated peptides may be substituted for the immunizing peptide to determine specificity of the antisera. Similarly, antisera raised against similar/related sequences may also be tested against the same set of peptides to further describe their selectivity.

4. Tissues should be frozen as soon as possible after surgery (for human tissues) or euthanasia (for animal tissues) and stored at –70°C to preserve tissue integrity

and minimize the potential for degradation of receptor protein and, hence, antigenicity. Care should be taken to snap-freeze tissue as quickly as possible to minimize the formation of ice crystals, which can lead to the formation of holes in the tissue. If tissues are cut into small blocks and quickly frozen, the need for repeated freeze-thaw cycles is reduced, the formation of ice crystals which may disrupt tissue morphology will be limited, and the smaller blocks will be less likely to crack during the freezing process. When storing tissues, care should be taken to wrap the tissue *loosely* in foil, to prevent it sticking. The appropriately labeled tissue can then be placed in a sealable bag to minimize dehydration, and then preferably into a box to prevent crushing, before storing at –70°C.

5. The cryostat chuck should be cooled either over dry ice or on the cold shelf of the cryostat; larger pieces of tissue may be mounted directly onto the chuck with mounting medium. For smaller tissues, a layer of mounting medium may be applied to the chuck and allowed to freeze before mounting of the tissue; this will allow a greater proportion of the tissue to be sectioned without the blade coming into contact with the chuck. Also, for smaller tissues, more than one piece of tissue may be mounted on the same chuck. In some laboratories, tissues are preferentially mounted onto cork discs for storage, and the discs mounted onto the cryostat chucks for sectioning and then removed. This is not recommended, as the inherent "sponginess" of the cork can lead to its compression during sectioning. This allows movement of the tissue while sectioning, producing inconsistencies in the thickness and quality of the sections obtained. In addition, the repeated freeze–thawing of the mounting medium used to "glue" the disc to the chuck may add to this variability—we have found that when mounting medium has been through a freeze–thaw cycle it fails to solidify adequately. In mounting tissues onto cryostat chucks, consideration should be given to the orientation of the tissue. For example, keeping the tissue parallel to the chuck surface will ensure that less trimming of the tissue is required before a complete section is achieved. Small tissues such as blood vessels may require support, e.g., that the vessel is held upright using fine forceps, until the mounting medium has frozen.
6. The temperature for optimal sectioning of a tissue will vary and should be determined empirically. When positioning the chuck on the tissue holder within the cryostat, the orientation of the tissue should be considered, depending on the shape of the tissue. An irregular piece of tissue, for example, would be better sectioned with the longest "side" rather than the irregular aspect or "point" as the leading edge, as this may result in the tissue section rotating as the section is cut, giving a creased or torn section. Good quality sections are imperative in interpreting results; sections of irregular thickness, or sections containing tears or folds, may result in entrapment of the antisera or edge effects, resulting in a falsely high signal being seen. Once the orientation is decided, it is a good idea to mark the chuck in order to enable subsequent sectioning of the same tissue to be achieved more easily and without loss of tissue resulting from trimming.
7. Cryostat microtomes are typically able to cut sections ranging from 5 to 30 μm in thickness. The optimum thickness is a balance between opposing factors. The

concentration of receptors tends to be less than other structural proteins. Thick sections (30 μm) may be required to detect receptors by ICC (compared with the 10 μm typically used in radioligand binding and autoradiography), but this can result in an increased background. Poly-L-lysine-coated slides are recommended to ensure adhesion of tissue during the long incubation steps and washing, particularly for large cryostat sections such as whole-body sections from mice.

8. We routinely use 12-well plates; this allows a combination of treatments to be performed (including duplicates, dilution ranges, negative controls [e.g., pre-immune sera, pre-absorption of antisera] and positive controls-antisera).
9. Cells are washed by tipping the plate at an angle and carefully aspirating off the incubation buffer from the bottom corner at the edge of the coverslip with a Pasteur pipet. Two milliliters PBS per well is then added then added slowly, with the pipet angled to the side of the well rather than directly onto the cells to prevent the cells lifting off. This is repeated one to two times. If cells are not to be used immediately, they may be fixed (1:1 mix of acetone:methanol) and allowed to dry before freezing at –80°C.
10. The purpose of fixing tissues is to preserve tissue morphology and retain receptor protein within the section during the staining process without altering the antigenicity. The strategy recommended here is that of postfixation of fresh-frozen sections by short exposure to acetone or formaldehyde. Acetone is a coagulating fixative that rapidly penetrates and reduces the volume of the tissue with some extraction of lipids. Formaldehyde does not coagulate proteins, but can modify the antigenic sequence; it has no effect on tissue volume or lipids. Therefore, testing both fixatives, which are comparatively mild is recommended. For some tissues, this may result in underfixation and a loss of morphology of the surrounding tissues, making interpretation of the staining difficult. If this is suspected, the length of fixation with formaldehyde can be increased, or tissue can be fixed in formaldehyde prior to sectioning or testing of other fixatives.
11. When manipulating slides, it is crucial to limit damage to the sections as much as possible. After each incubation and wash step, the excess buffer must be removed to prevent dilution of the next reagent. Care must be taken in wiping buffer from around the sections to prevent damage to the tissue. Using a hydrophobic pen to ring the tissue sections is recommended. It serves as a guide when wiping excess liquid from the slide, preventing the section from being accidentally wiped off, and by providing surface tension, it also allows smaller incubation volumes of antisera to be used without the solutions running off the slides. It is important not to allow the sections to dry out at any stage of the staining protocol, as this may result in uninterpretable results. Therefore, when processing a large number of slides, it is advisable to wipe only up to six slides at a time before adding the incubation solution. Pencil (or a diamond pen) should be used to label slides, as inks will be solubilized and lost during the alcohol dehydration and xylene stages.
12. For the fixation and wash stages, baths of reagents are the easiest and least labor-intensive method to use. However, for incubations with antisera, only small volumes of reagent are used as a result of the limitations of cost and availability of

reagents. The sections are therefore incubated horizontally with 100–300 µL incubation volume per slide, in incubation trays humidified to prevent the sections drying out.

13. The nonimmune or "normal" blocking serum chosen is usually from the same species that is providing the secondary antisera—i.e., if, as in this case, the second antisera is raised in swine, swine nonimmune sera is used. Following incubation, this reagent is not washed off but is tipped off; therefore, care must be taken to ensure that only a thin layer remains in order to prevent further dilution of the primary antisera.
14. The combination of antisera dilution, incubation time, and incubation temperature used are interdependent and will all have a bearing on the final nonspecific background staining and specific staining intensity achieved within a particular tissue. It is therefore necessary to determine the optimum conditions empirically. Commercial antisera are often supplied with suggested conditions for use; however, these may not necessarily be appropriate for every application. For positive staining but with high background, we recommend testing new primary antisera over a concentration range, e.g., 1:100, 1:200, 1:500, and 1:1000, for 24 h at 4°C while keeping the concentration of the other reagents constant. This strategy can usually reduce high backgrounds. Other alternatives are to use dilute primary antisera over a longer period, e.g., 24, 48, and 60 h at 4°C, to affinity-purify the antisera, if this has not already been done, to reduce the thickness of the section, to block endogenous peroxidase. For no positive staining, check the ICC protocol with another primary antisera from the same species known to work. Increase the concentration of the primary and/or secondary antisera, increase the incubation time up to 3 d, and/or incubate at room temperature or 37°C.
15. Because the antisera chosen to illustrate this protocol were raised in rabbits, the secondary antisera chosen for this system is swine antirabbit (at a dilution of 1:200). This provides the "link" between the primary antisera and the PAP complex, as both are raised in the same animal species.
16. The PAP complex (used at a 1:400 dilution) consists of the enzyme peroxidase and an antibody against peroxidase. The peroxidase complexes with the substrate, hydrogen peroxide (H_2O_2), and this in turn reacts with the chromogen (an electron donor) to produce a coloured product. It may be noted that in this reaction, the enzyme is not depleted and therefore, each molecule of enzyme bound is available to react with further hydrogen peroxide and subsequently produce more molecules of colored product, providing further amplification in signal. This is an advantage of the immunoperoxidase method over immunofluorescence techniques, in which one fluorescent molecule binds with no amplification step. Each visualization method has its own drawbacks; for example, the PAP method produces a permanent and intense staining, however endogenous peroxidase activity within certain tissues such as red blood cells, liver, kidney, and brain can produce nonspecific staining via a direct reaction with the H_2O_2 of the chromogen solution. Incubation of the sections with a methanolic H_2O_2 solution (1:4, mix of 3% H_2O_2:methanol) may be used to suppress the endogenous activity; however, this

treatment may also result in a decrease in adhesion of the section to the microscope slide, thus requiring the investigator to treat the sections more carefully during subsequent protocol steps.

17. The cultured cells should be manipulated gently to prevent them from lifting from the coverslips. A pair of forceps with curved fine tips may be used to lift each coverslip in turn from the wells and to dip them through the series of alcohol baths. Excess alcohol is removed by touching the edge of the coverslip onto absorbent paper, and the coverslip is lowered, cell side down, onto a pool of DePeX on a microscope slide.
18. Alexa Fluor 488 has a spectrum similar to the widely used fluorescein, but has a more intense fluorescent yield and better photo stability. Alexa Fluor 568 contrasts well with the 488 dye and is comparable with other dyes that fluoresce in the red (rhodamine and Texas Red) spectra.
19. Vectashield scavenges free radicals produced by excitation of fluorochromes and is added to nonpermanent mountants to reduce fading of the fluorescence signal, the major drawback of using immunofluorescent dyes. If immediate viewing is not possible, storage times may be increased by keeping slides wrapped in foil and refrigerated. If necessary, the sections can be re-stained.

Acknowledgments

We thank the British Heart Foundation for support.

References

1. Davenport, A. P. (2002) International Union of Pharmacology. XXIX. Update on endothelin receptor nomenclature. *Pharmacol. Rev.* **54,** 219–226.
2. Katugampola, S., and Davenport, A. P. (2003) Emerging roles for orphan G-protein-coupled receptors in the cardiovascular system. *Trends Pharmacol. Sci.* **24,** 30–35.
3. Katugampola, S. D., Pallikaros, Z., and Davenport, A. P. (2001) [^{125}I-His(9)]-ghrelin, a novel radioligand for localizing GHS orphan receptors in human and rat tissue: up-regulation of receptors with atherosclerosis. *Br. J. Pharmacol.* **134,** 143–149.
4. Davenport, A. P. (2003) Peptide and trace amine orphan receptors: prospects for new therapeutic targets. *Curr. Opin Pharmacol.* **3,** 127–134.
5. Davenport, A. P., and Macphee, C. H. (2003) Translating the human genome: Renaissance of cardiovascular receptor pharmacology. *Curr. Opin. Pharmacol.* **3,** 111–113.
6. Davenport, A. P. and Morton, A. J. (1991) Binding sites for ^{125}I ET-1, ET-2, ET-3 and vasoactive intestinal contractor are present in adult rat brain and neurone-enriched primary cultures of embryonic brain cells. *Brain Res.* **554,** 278–285.
7. Russell, F. D., Skepper, J. N., and Davenport, A. P. (1997) Detection of endothelin receptors in human coronary artery vascular smooth muscle cells but not endothelial cells by using electron microscope autoradiography. *J. Cardiovasc. Pharmacol.* **29,** 820–826.

8. Kuc, R.E. and Davenport, A. P. (2004) Comparison of ET_A and ET_B receptor distribution visualised by radioligand binding versus immunocytochemical localisation using sub-type selective antisera. *J. Cardiovasc. Pharmacol.* **43,** S224–S226.
9. Baker, S. J., Morris, J. L., and Gibbins, I. L. (2003) Cloning of a C-terminally truncated NK-1 receptor from guinea-pig nervous system. *Brain. Res. Mol. Brain Res.* **111,** 136–147.
10. Oliver, K. R., Wainwright, A., Edvinsson, L., Pickard, J. D., and Hill, R.G. (2002) Immunohistochemical localization of calcitonin receptor-like receptor and receptor activity-modifying proteins in the human cerebral vasculature. *J. Cereb. Blood Flow Metab.* **22,** 620–629.
11. Robas, N., O'Reilly, M., Katugampola, S., and Fidock, M. (2003) Maximizing serendipity: strategies for identifying ligands for orphan G-protein-coupled receptors. *Curr. Opin. Pharmacol.* **3,** 121–126.
12. D'Orleans-Juste, P., Honore, J.-C., Carrier, E., and Labonte, J. (2003) Cardiovascular diseases: new insights from knockout mice. *Curr. Opin. Pharmacol.* **3,** 181–185.
13. Kurihara, H., Kurihara, J., and Yazaki, Y. (2001) Lessons from gene deletion of endothelin systems. *Hdbk. Exp. Pharmacol.* **152,** 141–154.
14. Davenport, A. P., Kuc, R. E., Plumpton, C., Mockridge, J.W., Barker, P.J., and Huskisson, N. S. (1998). Endothelin-converting enzyme (ECE) in human tissue. *Histochem. J.*, **30**, 1–16.
15. Davenport, A. P. and Kuc, R. E. (2000) Cellular expression of isoforms of endothelin-converting enzyme-1 (ECE-1c, ECE-1b and ECE-1a) and endothelin-converting enzyme-2. *J. Cardiovasc. Pharmacol.* **36,** S12–S14.
16. Kuc, R. E. Immunocytochemical localization of endothelin peptides, precursors and endothelin-converting enzymes. *Methods Mol. Biol.* **206**, 3–9.

9

Live Cell Imaging of G Protein-Coupled Receptors

Burkhard Wiesner, Michael Beyermann, and Alexander Oksche

1. Introduction

In recent years, the endocytosis and the intracellular trafficking of many G protein-coupled receptors (GPCRs) have been evaluated. A milestone in the analysis of the transport of GPCRs was the molecular cloning of the green fluorescent protein (GFP) from the jellyfish *Aequorea victoria* by Prasher and coworkers ***(1,2)***. Site-directed mutagenesis yielded derivatives with higher photostability, improved quantum yield, and different excitation/emission spectra ***(3–5)***. For the generation of GPCR.GFP fusion proteins, the cDNA encoding GFP is in almost all cases genetically fused in frame to the 3'end of cDNA encoding the GPCR ***(6)***. The encoded fusion protein comprises a GPCR with the GFP moiety fused to its intracellular C-terminus. Although GFP is a relatively large protein (238 amino acids, 26.9 kDa) and its size is almost equal to that of most GPCRs (about 400 amino acids), the functional properties (ligand affinity, signal transduction, or intracellular trafficking) of GPCRs are not, or are only slightly, altered ***(6–8)***. Thus, GFP and its derivatives have been widely applied as fluorescent probes to visualize trafficking of GPCRs in real time, to analyze GPCRs' mobility by fluorescence recovery after photobleaching (FRAP; ***9***), and to study protein–protein interactions of GPCRs by fluorescence resonance energy transfer (FRET; ***10,11***). The relatively high stability of GFP and its chromophore in the presence of detergents and fixatives also allows the use of GPCR.GFP fusion proteins in co-localization studies with immunocytochemistry.

In contrast, GPCR fusion proteins with a red fluorescent fusion protein from the corallimorph *Discosoma* sp. (DSRed) were often found to be misrouted

From: *Methods in Molecular Biology, vol. 306: Receptor Binding Techniques: Second Edition*
Edited by: A. P. Davenport © Humana Press Inc., Totowa, NJ

within the cells. This missorting can most likely be attributed to the property of DsRed to form a tight tetramer with a nanomolar association constant *(12)*.

The use of GPCR.GFP fusion proteins in conjunction with fluorescent ligands allows the visualization of ligand–receptor interactions at a cellular level. In addition, the routes of intracellular trafficking for receptors and their ligands can be studied simultaneously *(13)*. Even when emission spectra of different fluorophores show considerable overlap, novel advances in laser scanning microscopy allow their spectral separation.

Despite the ongoing progress in the analysis of fluorescent samples, the resolution of light microscopy is limited to about 200 nm in the horizontal and to 100 nm in the vertical plane. Thus, the precise localization of a protein within a cell can only be determined by electron microscopy.

The methods described in the following paragraphs are based on the laser scanning microscopy (LSM)510 META system (Zeiss, Jena, Germany), but most of the protocols can also be performed with other LSM systems (from Biorad or Leica).

1.1. Co-Localization Studies

For the analysis of the subcellular distribution of wild-type and mutant GPCRs, staining of the plasma membrane and of intracellular compartments, e.g., the endoplasmic reticulum (ER) or the Golgi apparatus, is required. Identification of GPCRs within the plasma membrane can be difficult when cells have a flat morphology, as is the case for endothelial cells or smooth muscle cells.

The use of receptor-selective fluorescent ligands is the most convenient way to demonstrate the presence of a GPCR or a GPCR.GFP fusion protein at the plasma membrane *(13,14)*. However, commercially available fluorescent ligands are expensive. Alternatively, peptide ligands can be generated in variable quantities with different dyes commercially available from Amersham Biosciences (Cyanin dyes) or Molecular Probes (Alexa dyes). However, the fluorescent peptides have to be tested by high-performance liquid chromatography (HPLC) and mass spectometry (MS) for dye incorporation and in ligand-binding studies or in signal transduction analysis for their functionality. A protocol for the synthesis and labeling of peptide ligands is given in **Subheading 3.4.1.**

If receptor-selective ligands are not available, plasma membrane localization can be verified experimentally by studies with trypan blue or the cationic styrylpyridinium dye FM 1-43, which both specifically label the plasma membrane. Alternatively, GPCR. cyan fluorescent protein (CFP) or GPCR. yellow fluorescent protein (YFP) proteins can be co-expressed with plasma membrane-targeted Mem.YFP or Mem.CFP fusion proteins, respectively (*see* **Table 1**; available from Becton Dickinson).

Table 1
Organelle- and Plasma Membrane-Targeted Green Fluorescent Protein (GFP) Fusion Proteins

	Targeting sequence	Localization
pEGFP.Endo	RhoB, human	early/late endosomes
pEGFP.Golgi	N-terminal 81 aa of β1,4 glactosyltransferase	trans/medial Golgi
pEGFP.ER.	N-terminal signal peptide of calreticulin and C-terminal KDEL sequence	ER
pEGFP.Nuc	Three tandem copies of the SV40 T-antigen nuclear localization signal	Nucleus
pEGFP.Mem	N-terminal 20 aa of neuromodulin	Plasma membrane
pEGFP.Mito	Cytochrome C oxidase, precursor of subunit VIII, human	Mitochondria
pEGFP.Actin	β-actin, human	Actin
pEGFP.Tub	α-tubulin	Tubulin

The table summarizes several different, commercially available GFP fusion proteins, which are sorted to the plasma membrane, organelles, or cytoskeletal proteins. The sequences mediating targeting are noted.

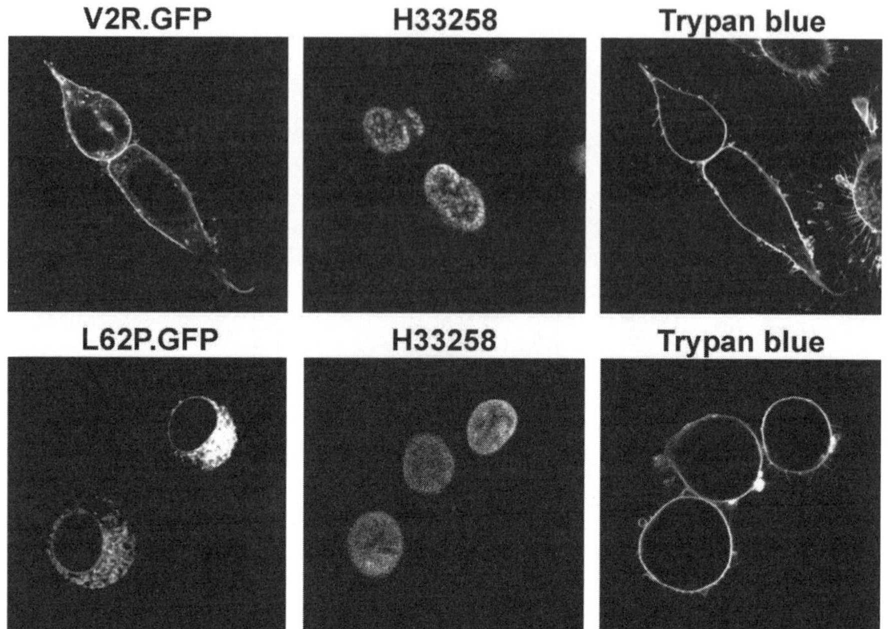

Fig. 1. Analysis of cell-surface expression of GPCR.GFP fusion proteins. HEK293 cells transiently expressing the wild-type (upper panel) and the L62P mutant (lower panel) human V_2R.GFP fusion protein were analyzed by laser scanning microscopy. ***Left panel***: V_2R.GFP fusion proteins. ***Middle panel***: Staining of nuclei with H33258. ***Right panel***: Staining of the plasma membrane with trypan blue.

In the case of GPCR.GFP fusion proteins that are localized within intracellular compartments as a result of improper folding (retention in the ER), missorting as a result of altered targeting signals (e.g., to lysosomes), or as a result of agonist-induced internalization, the intracellular organelles can be assigned by different markers, which are either added to the medium or co-expressed with the GPCR.GFP fusion protein. A comprehensive list of membrane-permeable markers for the ER, the Golgi apparatus, lysosomes, mitochondria, and the nucleus is found in the Molecular Probes Catalog (www.probes.com). Various organelle-targeted CFP/GFP/YFP fusion proteins, e.g., for ER- or Golgi-targeted fusion proteins, are commercially available from Becton Dickinson (http://www.bdbiosciences.com/clontech/gfp/index.shtml). Protocols for co-localization experiments of GPCR.GFP proteins with plasma membrane and organelle-specific markers are presented in **Subheading 3.3.**

In **Fig. 1**, the subcellular distribution of a wild-type and a mutant V_2R.GFP is shown. The outline of the plasma membrane is visualized with trypan blue and the nucleus is stained with the membrane-permeable bisbenzimide dye

H33258 (Molecular Probes). Although the pattern of distribution observed for the wild-type V_2R.GFP is similar to the pattern obtained with trypan blue, the mutant V_2R.GFP is mainly found intracellularly, not within the nucleus.

1.2. Ligand-Induced Internalization

When GPCR.GFP fusion proteins are expressed in native cells or cell lines, the endocytosis and intracellular trafficking of the agonist-bound receptor can be monitored at the level of a single cell. In conjunction with fluorescent agonists, it is even possible to study ligand–receptor interactions directly. In **Fig. 2**, an example of such a time-lapse experiment is shown. Cells stably expressing an endothelin A (ET_A) receptor GFP fusion protein were incubated with Cy3-ET1 and binding of the ligand to the ET_A.GFP at the plasma membrane was recorded every 11 s. On the basis of these images, the time required for the ligand to bind to the receptor can be calculated.

The ligand-induced internalization can be monitored in real time, and an example is shown in **Fig. 3**. With the use of pH-sensitive dyes, such as CypHer5 from Amersham Biosciences, which is fluorescent only in an acidic environment, it is even possible to visualize only those receptors already transported to acidic compartments, such as early and late endosomes. Further, it is possible to study the mode of receptor internalization (clathrin- or caveolae-mediated endocytosis) and to determine the intracellular trafficking routes of the ligand-bound GPCRs. This involves the use of chemicals and drugs or the co-expression of proteins, which interfere with the internalization or the further transport of a GPCR. Protocols for the analysis of the agonist-induced endocytosis and the intracellular trafficking routes are presented in **Subheading 3.4**.

It is also possible to visualize the proteins involved in the internalization of GPCRs, e.g., kinases and adaptor proteins directly. A key step in the regulation of activated GPCRs is their desensitization by protein kinases. In most cases, the desensitization of the agonist-bound GPCR is initiated via GPCR kinases (GRK), which phosphorylate the receptors at serine/threonine residues in the C-terminus and/or the third intracellular loop ***(15,16)***. The phosphorylated GPCRs have a high affinity to visual arrestin (in the case of rhodopsin) or isoforms of β-arrestins (β-arrestin1 and β-arrestin2 in the case of nonvisual GPCRs; ***17***). Binding of arrestin or β-arrestin isoforms to the GPCR prevent further interaction with G proteins and, in addition, initiates the receptors internalization ***(18–20)***.

By the combined expression of GRK.DsRed and GPCR.GFP fusion proteins, it is also possible to monitor GRK recruitment to the plasma membrane following receptor activation ***(21)***. Similarly, the activation-dependent recruitment of arrestin or β-arrestins to GPCRs can be visualized by co-expression of GPCRs with arrestin.GFP or β-arrestin1/2.GFP fusion proteins ***(18,22)***.

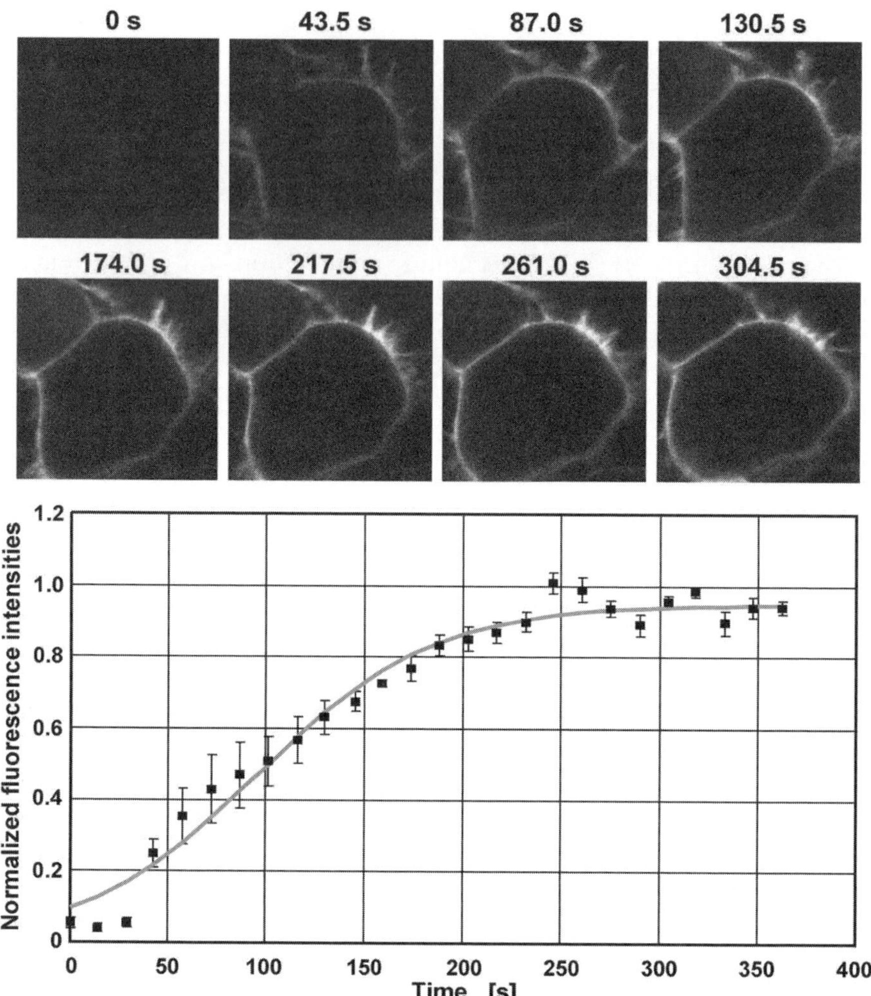

Fig. 2. Laser scanning microscopy (LSM) analysis of a ligand/G protein-coupled receptor (GPCR) interaction. HEK293 cells stably expressing a fusion protein (ET_A.GFP) comprising of the ET_A receptor and the green fluorescent protein (GFP) were incubated with the fluorescent ligand Cy3-ET-1. Cells were placed in a temperable insert on the LSM510 META and incubated for up to 6 min at 37°C. With the application of Cy3-ET-1, a time series was started and images were recorded every 11 s. ***Upper two panels***: LSM images of the time-lapse series are shown (every fourth image). ***Bottom panel***: graphical presentation of the time-dependent increase in the fluorescence at the plasma membrane (as normalized fluorescence intensities). The values are means of four different regions of interest set at the plasma membrane ± SEM. The half maximal time required for the association of the ligand to the receptor is 96.7 ± 4.3 s.

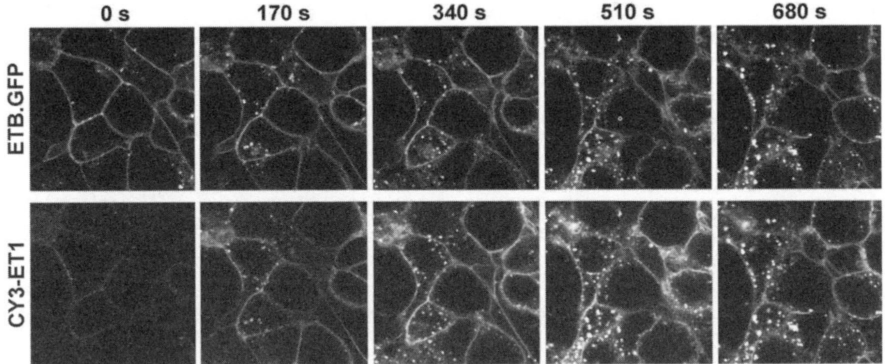

Fig. 3. Real-time analysis of agonist/G protein-coupled receptor (GPCR) internalization. HEK293 cells stably expressing a fusion protein (ET_B.GFP) comprising the endothelin B (ET_B) receptor and the green fluorescent protein (GFP) were stimulated with the fluorescent ligand Cy3-ET-1. Cells were placed in a temperable insert on the LSM510 META and incubated for up to 12 min at 37°C. A time series was started with the addition of the fluorescent ligand, and images were recorded every 42 s (every fourth image is shown). ***Upper panel***: subcellular distribution of the ET_B.GFP fusion protein. ***Lower panel***: subcellular distribution of Cy3-ET-1.

1.3. Transport to the Plasma Membrane

GFP fusion proteins can also be used to study the transport of GPCRs from the ER via the Golgi to the plasma membrane. Following synthesis and proper folding within the ER, GPCRs are transported to the Golgi apparatus. Here, posttranslational modifications, e.g., *N*- and *O*-linked glycosylation, and palmitoylation, occur. The properly folded, complex-glycosylated GPCR is then targeted to the plasma membrane. However, when the GPCR fails to fold properly, it remains retained in the ER and finally becomes degraded via the proteosome.

There are several conditions in which GPCRs show inefficient folding and maturation. This may be caused by an alteration in the amino acid sequence (mutation), or is the result of a low conformational stability of the wild-type protein. For some ER-retained GPCRs, it has been shown that specific membrane-permeable antagonists can promote proper folding, thereby enabling the exit of the GPCR from the ER *(23–26)*. This antagonist-mediated restoration of cell surface expression of ER-retained mutant GPCRs can be analyzed and quantified by LSM studies. A protocol for such a quantitative LSM analysis is presented in **Subheading 3.5.**

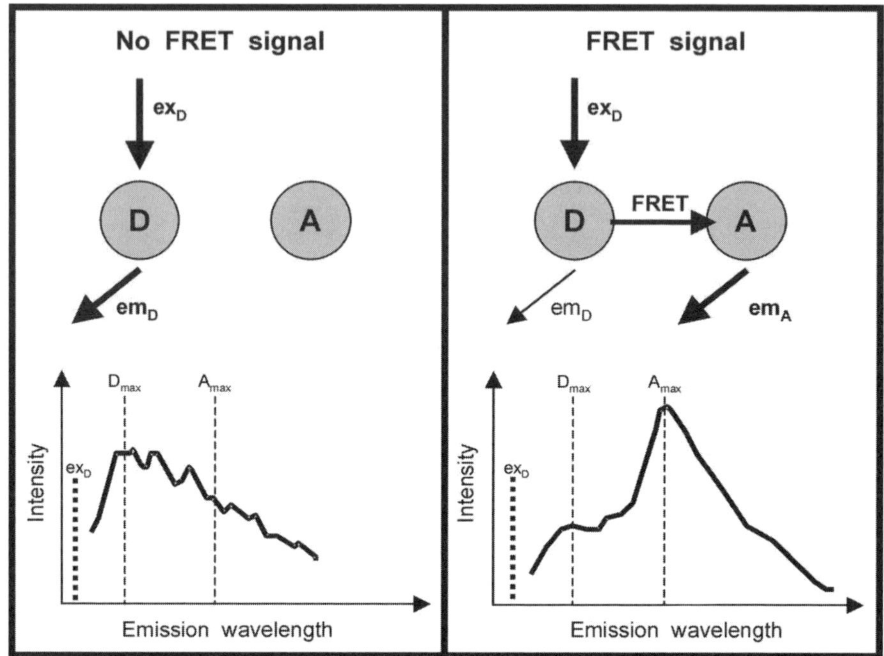

Fig. 4. The principle of the fluorescence resonance energy transfer. Donor cyan fluorescent protein (CFP) and acceptor yellow fluorescent protein (YFP) fluorophores are co-expressed in a single cell. When no interaction of CFP with YFP occurs (**A**), excitation of CFP results in an emission spectrum representative for CFP. When CFP and YFP interact with each other and CFP is excited (**B**), energy is transferred from CFP to YFP resulting in a reduced CFP and an increased YFP emission. Ex_D, excitation of the donor; em_D, emission of the donor; em_A, emission of the acceptor; D_{max}, maximum of the fluorescence of the donor; A_{max}, maximum of the fluorescence of the acceptor; FRET, fluorescence resonance energy transfer.

1.4. Fluorescence Resonance Energy Transfer

FRET is a nonradiative transfer of photon energy from a donor to an acceptor fluorophore. This energy transfer occurs only when both fluorophores are in close proximity (less than 10 nm) and is thus indicative of a direct protein–protein interaction. **Figure 4** schematically describes the basis of FRET. FRET can be used to monitor dynamic or static interactions between GPCRs such as ligand-induced or constitutive oligomerization of GPCRs *(**10,11**)*. It is also possible to monitor structural changes within a GPCR when donor and acceptor fluorophores are inserted into the third intracellular loop and the C terminus of a GPCR, respectively *(27)*. Thereby, the kinetics of receptor–ligand interactions can be analyzed.

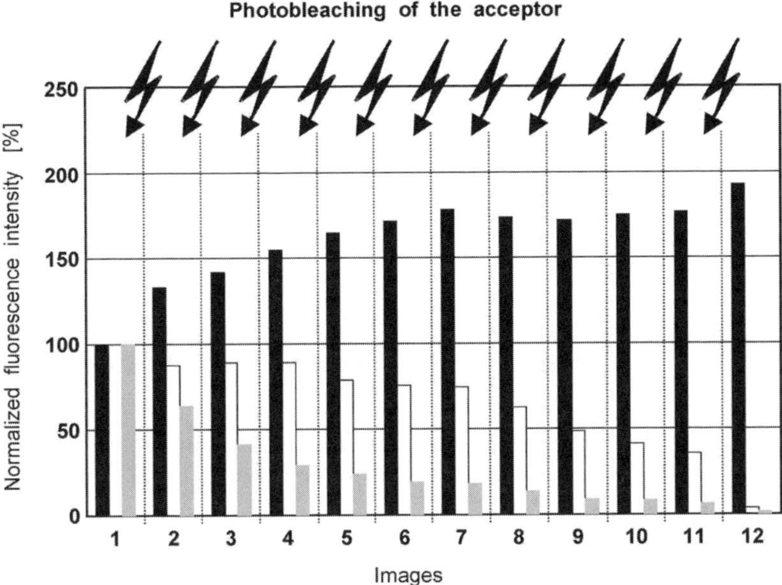

Fig. 5. Fluorescence resonance energy transfer (FRET) analysis. HEK293 cells transiently expressing the cortisol releasing factor (CRF) receptor as cyan fluorescent protein (CFP) and yellow fluorescent (YFP) fusion protein were analyzed in the laser scanning microscopy (LSM)510 META system using the "Multi-track" mode. Three channel images (CFP, YFP, FRET) were taken before and after each bleaching period of the acceptor (YFP). Bleaching (indicated by a thunderbolt) was performed in 11 cycles. The fluorescence intensities obtained for CFP (black), YFP (white), and FRET (grey) of each image were normalized to the fluorescence intensity obtained for CFP, YFP, and FRET before photobleaching.

In the case of dynamic protein–protein interactions, an increase and decrease in FRET will be observed, which is in most cases sufficient to demonstrate the existence of a regulated protein–protein interaction. However, for static protein–protein interactions, it is crucial to exclude false–positive FRET values. This can be done by photobleaching the acceptor, which results in an increase of the donor emission only when a protein–protein interaction exists. In **Fig. 5,** FRET analysis with photobleaching of the acceptor reveals homodimerization of corticotropin releasing factor receptor type 1 (CRF-1) CFP and YFP fusion proteins (Krätke et al., unpublished observation). In addition, by using the photobleaching FRET approach, the FRET efficiency can be determined. Because the FRET efficiency depends on the level of protein–protein interaction, it can be used to determine the regulation of dimerization/oligomerization under various experimental conditions. A protocol for a FRET analysis with photobleaching of the acceptor is presented in **Subheading 3.6.**

2. Materials

2.1. Transient and Stable Expression of GFP Fusion Proteins in Mammalian Cells

2.1.1. Coverslips

1. Round coverslips (thickness between 125 and 170 μm; preferred diameter is 24–30 mm).
2. Poly-L-lysine as dry substance (P-5899, Sigma, Munich, Germany) or as sterile solution (100 μg/mL; L 7240, Biochrom, Berlin, Germany; storage for several months at 4°C).

2.1.2. Holder for Coverslips and Heating Unit for Motorized or Manual Xy-Stage

1. Holder for coverslips are commercially available from PeCon (Erbach, Germany; also distributed by Zeiss, Jena, Germany) or can be fabricated according to **Fig. 6**.
2. Temperable insert and heating unit (**Fig. 6**; available from Zeiss).
3. Optional heating/CO_2 incubation chamber for Zeiss microscope M200, available from PeCon (Erbach, Germany; also distributed by Zeiss, Jena, Germany).

2.1.3. HEPES-Buffered Media

1. Dulbecco's modified Eagle medium (DMEM), phenol red-free (Sigma). Add 1 m*M* L-glutamine before use and store at 4°C. Use within 2 wk.
2. HEPES as dry substance (Sigma) or as sterile solution (1 *M*, adjusted to pH 7.4, Amersham Biosciences, Braunschweig, Germany).
3. Krebs-Ringer HEPES buffer (KRH): 125 m*M* NaCl, 3 m*M* KCl, 1 mM NaH_2PO_4, 1.2 m*M* $MgSO_4$, 2.4 m*M* $CaCl_2$, 22 m*M* $NaHCO_3$, 5.5 m*M* glucose, 10 m*M* HEPES.
4. Dulbecco's phosphate-buffered saline solution (DPBS): 137 m*M* NaCl, 2.7 m*M* KCl, 1.5 m*M* KH_2PO_4, 8.0 m*M* Na_2HPO_4, 0.9 m*M* $CaCl_2$, 0.5 m*M* $MgCl_2$, pH 7.4.

2.2. Co-Localization Analysis

For the identification of the plasma membrane and cellular organelles in living cells, several fixable and nonfixable marker proteins are commercially available. A comprehensive list of chemicals is given in the Molecular Probes Catalog. Most of the commercially available plasma membrane- and organelle-targeted GFP fusion proteins are available from Becton Dickinson and are summarized in the Becton Dickinson Catalog.

2.2.1. Plasma Membrane

For the identification of the plasma membrane in living cells, we use the following markers:

1. *Chemicals*: trypan blue (Sigma) and FM 1-43 (Molecular Probes, Leiden, The Netherlands) specifically label the plasma membrane.
2. *Fusion proteins for co-expression analysis*: CFP, GFP, and YFP are available as fusion proteins with a plasma membrane targeting sequence (Mem.CFP/GFP/YFP; commercially available from BD, Heidelberg, Germany; **Table 1**).

2.2.2. Cellular Organelles

For the identification of cellular organelles in living cells, we use the following markers:

1. *Chemicals*: rhodamine 6G chloride (ER, mitochondria), fluorescent ceramide (Golgi apparatus), Mitotracker (mitochondria), and the bisbenzimide dye H33258 (nucleus; all from Molecular Probes).
2. *Fusion proteins for co-expression analysis*: CFP, GFP, and YFP are available as fusion proteins with targeting sequences for the ER, Golgi apparatus, mitochondria and nucleus, as well as for cytoskeletal proteins (*see* **Table 1**).

2.2.3. Markers of Endocytosis

For the identification of the different routes of internalization in living cells, we use the following markers:

1. *Chemicals*: lysotracker and fluorescent dextran (late endosomes/lysosomes; Molecular Probes).
2. *Fusion proteins for co-expression analysis*: CFP, GFP, and YFP fusion protein of Rho B can be used (Becton Dickinson).
3. *Fluorescent marker proteins*: fluorescent transferrin (early endosomes, percentriolar recycling compartment) or fluorescent low-density lipoprotein (LDL) (late endosomes, lysosomes) are available from Molecular Probes (Leiden, The Netherlands).
4. *Other chemicals*: desferoxamine (Sigma; final concentration of 4 μ*M*), 450 m*M* sucrose in DMEM, or KRH buffer.

2.3. Labeling of Peptides With Fluorochromes

1. *N*-α-9-fluorenylmethyloxycarbonyl (Fmoc)-amino acids (Novabiochem AG, Germany).
2. Dimethyl sulfoxide (DMSO) (Sigma).
3. Lyophilizer, preparative HPLC with a C18 Nucleosil column, matrix-assisted laser desorption/ionization (MALDI)-MS mass spectrometer.
4. Analytical HPLC equipped with ultraviolet (UV) and fluorescence detectors.

2.3.1. Dyes

Fluorochromes for the labeling of free amino or thiol groups are commercially available from different companies, e.g., Cyanin dyes (Amersham Biosciences; http://www1.amershambiosciences.com) or Alexa dyes (Molecular

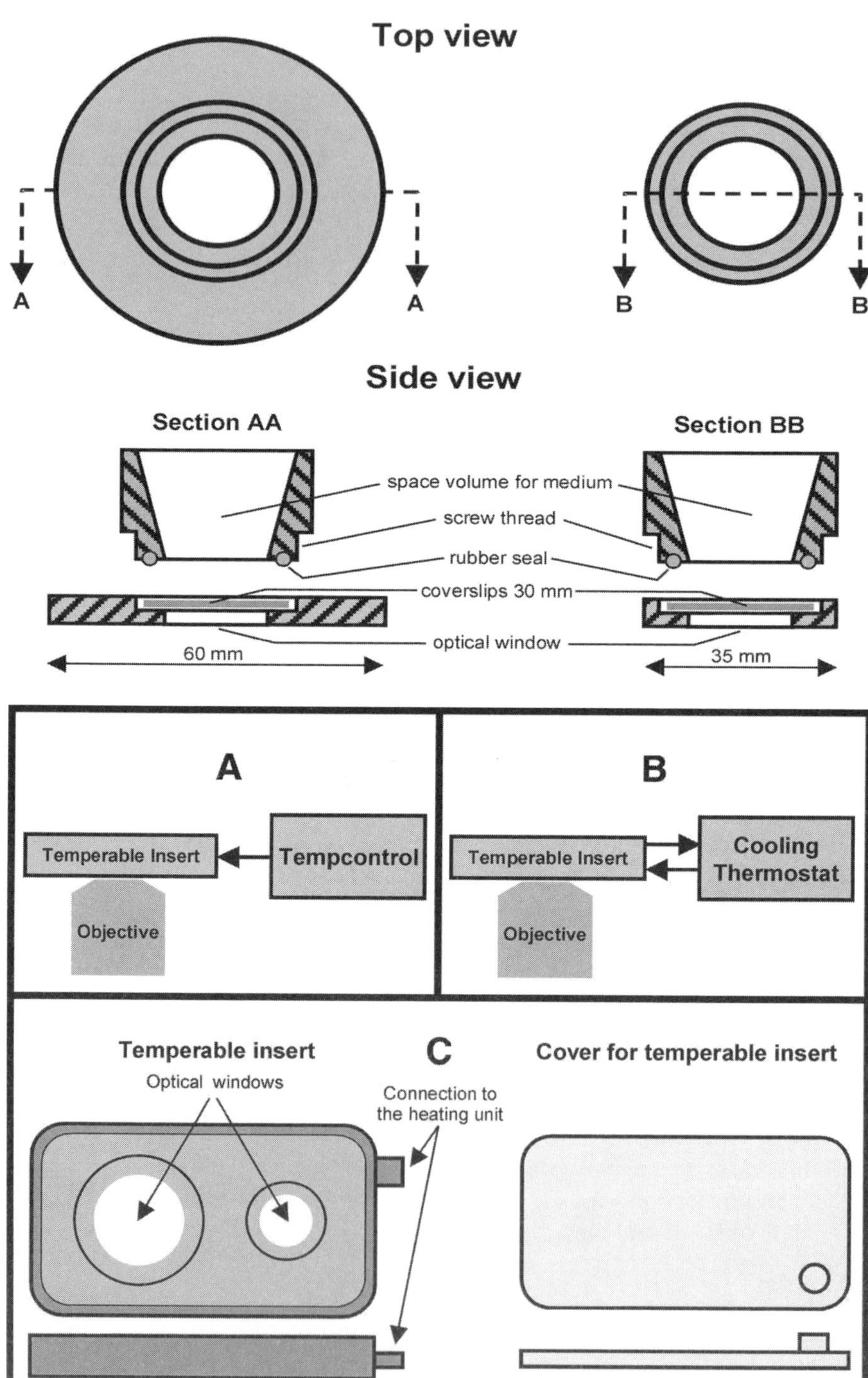
Top view
A
A
B
B
Side view
Section AA
Section BB
space volume for medium
screw thread
rubber seal
coverslips 30 mm
optical window
60 mm
35 mm
A
Temperable Insert
Tempcontrol
Objective
B
Temperable Insert
Cooling Thermostat
Objective
Temperable insert
Optical windows
C
Cover for temperable insert
Connection to the heating unit

Probes http://www.probes.com). For the labeling of peptide ligands, we use monoreactive *N*-hydroxysuccinimide (NHS) esters of Cyanin dyes (labeling of free amino groups).

1. NHS ester of Cy3, Cy3.5, Cy5, Cy5.5, or Cy7 (available in pack sizes of 1–50 mg).
2. NHS ester of CypHer5 can be used to create a pH-sensitive peptide ligand. CypHer5 is nonfluorescent at neutral pH, but fluorescent at acidic pH.

2.3.2. Labeling Procedure and Ligand Purification

1. 0.6 *M* Tris-HCl buffer, pH 6.8.
2. 0.1 *M* $NaHCO_3$, pH 8.5, and 0.1 *M* $NaHCO_3$, pH 9.3 (adjusted with NaOH).
3. Acetonitrile/dioxane (1:1, v/v).
4. *N,N*-diisopropylethylamine (Sigma).

3. Methods

3.1. Transient and Stable Expression of GFP Fusion Proteins in Mammalian Cells

3.1.1. Coverslips

1. Autoclave glass coverslips.
2. Place dry coverslips in a Petri dish and coat coverslip with poly-L-lysine (20 µg/mL) for 20 min at room temperature (*see* **Note 1**).
3. Remove poly-L-lysine solution and let coverslips dry.
4. Seed cells onto coverslips (50,000–100,000 cells/35 mm dish; *see* **Note 2**).
5. Force coverslips to the bottom of the petri dish with a sterile syringe or a pair of forceps (*see* **Note 1**).
6. Transfect cells the following day (in the case of transient transfection; *see* **Note 3**).

3.1.2. Holder for Coverslips for Motorized or Manual xy*-Stage*

1. Transfer coverslips to a holder, cover cells with KRH, and make sure that no fluid leaks out (*see* **Note 4**).
2. Remove any remaining fluid from the bottom of the coverslips.
3. Add a small drop of oil onto objective (inverted microscope). Use objectives with high magnification, e.g., 63× or 100×.

Fig. 6. *(continued from previous page)* Construction plan of a coverslip holder for the use in temperable inserts. (**I**), Shown are top and side views of two different coverslip holders which fit on the standard *xy*-stage of microscopes. The holders can also be used for temperable inserts (**IIC**) which fit on mechanical and motorized scanning stages. The temperable inserts can be used with electrical heating (temperature range: 3°C above room temperature to 60°C; **IIA**) or with a thermostat (temperature range: 0°C to 65°C; **IIB**).

4. Place holder with coverslip in a temperable insert, fixed at the microscopical stage (*see* **Note 5**).

3.2. Laser Scanning Microscopy

3.2.1. Laser Scanning Microscopy: Equipment and Settings

1. For the analysis of GFP and its spectral variants, argon lasers can be used. The laser lines are 458 nm (for CFP), 488 nm (for GFP), and 514 nm (for YFP; *see* **Note 6**).
2. For the analysis of fluorophores such as FITC, Cy2, or Alexa 488, the same laser line as for GFP is used (488 nm).
3. For red fluorophores such as TRITC, Cy3, or Cy3.5, a HeNe laser with a laser line at 543 nm is required.
4. For far red fluorophores such as Cy5 or Cy5.5, a HeNe laser with a laser line at 633 nm is required.
5. For the use of chemicals, which require excitation by UV (e.g., the bisbenzimide dye H33258), an additional excitation source, e.g., an argon UV laser with laser lines at 351 nm and 364 nm, is needed. The use of a violet diode (405 nm) or of an ionizing radiation (IR)-laser (720–930 nm, two-photon microscopy) is also possible.
6. For the simultaneous measurement of two or more dyes, choose separate detectors (photomultipliers). The scanning mode can be set to "Frame mode" if the protein movements are rather slow (>1 s) and to "Line mode" if kinetics are fast (ms; *see* **Note 7**).
7. A simple guide of how to set excitation wavelengths and emission windows is given in **Table 2**. The settings apply for the analysis of the different fluorophores in the "Multi-track" mode. For the analysis of multiple fluorophores with overlapping emission spectra, the use of the linear unmixing mode is recommended (*see* **Subheading 3.2.2.**).

3.2.2. Linear Unmixing in Multicolor Labeling Studies

1. For a variety of fluorophores, a significant overlap exists so that reliable multichannel measurements are not possible. For example, CFP, GFP, and YFP have grossly overlapping spectra, which cannot be resolved by LSM in the "Multi-track" mode. For these applications, a linear unmixing function is required on the basis of an emission fingerprint of a single fluorophore. The linear unmixing option is offered with the LSM510 META system (Zeiss).
2. The analysis of samples labeled with different fluorophores and overlapping emission spectra can be accomplished in a three-step procedure.
3. As a first step, the spectral signature of each fluorophore has to be determined separately. A lambda stack of images is scanned for each fluorophore and stored in the dye data base (*see* **Note 8**).
4. In a second step, a lambda stack of images from the sample (labeled with different fluorophores) is recorded.

Table 2
Laser Scanning Mikcroscopy (LSM) 510 META Settings for the Analysis of Different Fluorophores

Mode	Fluorophore	Excitation wavelength	Main beam splitter (HFT)	Beam splitter (NFT)		Channel	Emission filter
Line	H33258	405 nm	405/488/543	490	- - -	1	BP420-480
	GFP	488 nm	405/488/543	490	- - -	2	LP505
Line	GFP	488 nm	488/543	545	- - -	1	BP505-550
	Trypan blue	543 nm	488/543	545	- - -	2	LP560
Line	GFP	488 nm	488/543	545	- - -	1	BP505-530
	Rhodamine 6G	543 nm	488/543	545	- - -	2	LP560
Line	GFP	488 nm	488/543	545	- - -	1	BP505-530
	CY3	543 nm	488/543	545	- - -	2	LP560
Line	CFP	405 nm	405/488/543	545	490	1	BP420-480
	GFP	488 nm	405/488/543	545	490	2	BP505-530
	CY3	543 nm	405/488/543	545	- - -	3	LP560
Line	GFP	488 nm	UV/488/543/633	635VIS	- - -	1	LP505
	CY5	633 nm	UV/488/543/633	635VIS	- - -	2	LP650
Line	H33258	364 nm	UV/488/543/633	635VIS	490	1	BP385-470
	GFP	488 nm	UV/488/543/633	635VIS	490	2	LP505
	CY5	633 nm	UV/488/543/633	635VIS	- - -	3	LP650
Frame	H33258	364 nm	UV(375)	545	- - -	1	BP385-470
	YFP	514 nm	458/514	545	- - -	2	BP530-600
	CY5	633 nm	UV/488/543/633	635VIS	- - -	3	LP650
Frame	CFP	405 nm	405/514	635VIS	490	1	LP420-480
	YFP	514 nm	405/514	635VIS	545	2	BP535-590
	CY5	633 nm	UV/488/543/633	635VIS	- - -	3	LP650

The table summarizes the excitation wavelengths, beam splitters, and emission filters for the simultaneous analysis of different fluorophores in the "Multi-track" mode, which can be used for the live cell imaging of G protein-coupled receptors. GFP, green fluorescent protein; CFP, cyan fluorescent protein; YFP, yellow fluorescent protein.

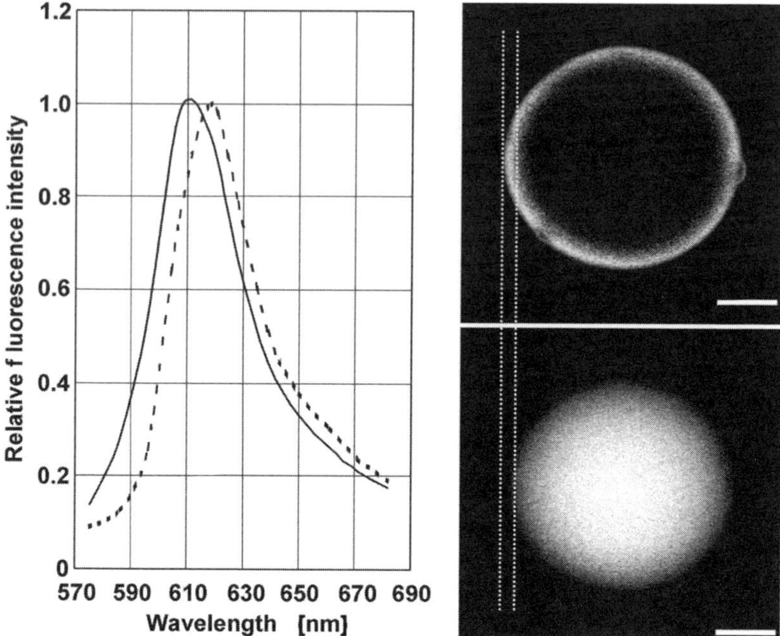

Fig. 7. Emission fingerprinting of a fluorescent sample. A bead labeled with two slightly different fluorophores, one on the inside and the other in the coat, was analyzed with the laser scanning microscopy (LSM)510 META system using "Emission Fingerprinting." Despite the highly overlapping emission spectra, the fluorophores within the bead's interior and the coat can be separated clearly. Bar: 2.5 µm.

5. In the final step the mode "Linear unmixing" is activated. The spectral signatures of all fluorophores represented in the sample are imported from the dye data base and the analysis is started.
6. Linear unmixing allows the spectral separation of two fluorophores, even when their emission maxima differ only by few nanometers. In **Fig. 7**, an example for such a spectral unmixing is shown.

3.3. Analysis of the Subcellular Distribution of GPCR.GFP Fusion Proteins

- To determine the subcellular localization of a GPCR.GFP fusion protein by LSM, simultaneous analysis of the GPCR.GFP fusion protein and of the plasma membrane or of different organelles is required.
- For live cell imaging, two approaches are possible: (a) live cell markers added to the medium stain the plasma membrane or cellular organelles, or (b) fusion proteins with defined targeting sequences directing CFP, GFP, or YFP to the plasma membrane or cellular organelles.

- *Live cell markers*: place the coverslip in the holder and cover cells with 500 μL medium. For visualization of the plasma membrane, add 20 μL of trypan blue (0.1% in PBS) to the region of interest or incubate cells with FM 1-43 (at a final concentration of 5 μ*M*). For the visualization of the nucleus, add the bisbenzimide dye H33258 (final concentration of 10 μg/mL), for the ER and mitochondria, add rhodamine 6 G chloride (final concentration of 5 μ*M*), and for the Golgi, fluorescent ceramide (final concentration of 5 μ*M*; *see* **Note 9**).
- *GFP fusion proteins for the identification of the plasma membrane and organelles*: for co-localization studies of GPCRs with CFP, GFP, or YFP moieties and plasma membrane or organelle-directed CFP, GFP, or YFP proteins, cells have to be co-transfected (*see also* **Note 5**). A variety of organelle-targeted GFP fusion proteins are commercially available from Becton Dickinson (*see* **Table 1**).

3.4. Analysis of the Endocytosis of Fusion Proteins

3.4.1. Generation of Fluorescent Ligands

1. Labeling of peptide ligands with Cyanin- or Alexa-based NHS esters can be performed with rather small quantities of peptides (100–500 μg). However, for several reasons, the labeling of larger quantities of peptides is recommended (*see* **Note 10**). Peptides can be synthesized by solid phase synthesis using the Fmoc-strategy (a review of Fmoc-based peptide synthesis is described in the Novabiochem Catalog).
2. Following acidic deprotection and cleavage of the synthesized peptide from the resin, the crude linear peptide is purified by preparative HPLC.
3. In the case of peptides with disulphide bonds (e.g., vasopressin, urotensin II, or endothelins), cyclization (disulphide bond formation) is required. Linear peptides (1 mg/mL) are dissolved in sodium bicarbonate buffer, pH 9.3, and exposed to air for 2–3 d under continuous stirring. Addition of 10–15 vol% DMSO can accelerate the cyclization.
4. Concentrate the peptide by lyophilization and purify it by preparative HPLC.
5. Characterize the peptide by mass spectrometry with MALDI- or electrospray ionization (ESI)-MS.
6. For labeling at the N-terminal α-amino group, the peptide is dissolved in 100 μL of 0.6 *M* Tris-HCl buffer, pH 6.8 (final concentration 30 μ*M*). For labeling of peptides at the ε-amino group of lysine residues, the peptide is dissolved in 100 μL of sodium bicarbonate buffer, pH 8.5 (final concentration 30 μ*M*).
7. Dissolve Cy-based NHS-ester (5.25 μ*M* final concentration) in 400 μL acetonitrile/dioxane (*see* **Note 11**).
8. Mix the peptide (100 μL) and NHS ester (400 μL) and start the labeling reaction by addition of *N,N*-diisopropylethylamine (10 μL). Control pH of the mixture. For the labeling of the *N*-terminal α-amino group, the pH should be in the range of 7.0–8.0, for the labeling of the ε-amino group of lysine, the pH should be around of 9.3.
9. Incubate the mixture overnight at room temperature and protect it from light (*see* **Note 12**).

10. Separate the labeled peptide from the nonlabeled peptide by preparative HPLC.
11. Control incorporation of the dye by analytical HPLC and verify the position of dye incorporation by MS with MS/MS fragmentation. Analyze the functional activity of the fluorescent ligands (*see* **Note 13**).

3.4.2. Analysis of the Agonist-Induced Internalization

1. Place the coverslip with cells expressing the GPCR.GFP fusion protein in the holder and cover cells with 250 μL medium.
2. Place the holder into the temperable insert (*see* **Note 14**).
3. Choose the "Multi-track" mode and set scan to "Line mode." Set channels according to the list in **Table 2**.
4. Identify cells expressing GPCR.GFP fusion proteins, which are convenient for LSM, and adjust the settings. Also activate the channel for the fluorescent ligand (e.g., Cy3). Here, the setting should be adjusted to parameters, for which clear signals were obtained in a previous experiment.
5. Activate the menu "Time series." In the menu, set the time period required to analyze the cells and choose the number of scans to be taken in this period (e.g., 20 scans with intervals of 1 min; *see* **Note 15**).
6. Start the time series and record three scans in the absence of the fluorescent ligand (baseline recording).
7. Add 250 μL of a prewarmed solution with the fluorescent ligand to the cells. For most GPCRs with nanomolar affinities, 10–100 n*M* of the fluorescent ligand as final concentration are sufficient.
8. After the time series has stopped, the images are stored in a gallery (*see* **Note 16**).

3.4.3. Characterization of Endocytic Pathways and Intracellular Trafficking

1. Clathrin-mediated endocytosis can be inhibited when sucrose (450 m*M*) is added to the medium prior to the addition of the agonist. When using a fluorescent agonist, it can also be demonstrated that binding of the agonist is not affected by the addition of 450 m*M* sucrose to the medium.
2. Although internalization via caveolae is not affected by the addition of sucrose (450 m*M*), co-expression of a dominant-negative mutant of dynamin I (K44A.dynamin I) abrogates clathrin- and caveolae-mediated endocytosis *(28)*.
3. The intracellular trafficking routes can be assigned best by the analysis of the internalization of the GPCR in combination with that of fluorescent transferrin (marker for recycling) or fluorescent LDL (marker for late endosomal and lysosomal transport).
4. For co-incubation with fluorescent transferrin, cells should be pretreated for 24 h with desferoxamine mesylate (4 μ*M*), which chelates iron. This pretreatment enhances uptake of fluorescent transferrin to the cells.
5. For co-incubation with fluorescent LDL, cells should be serum-starved for at least 24 h. This pretreatment results in an increased LDL receptor expression enhances the uptake of fluorescent LDL.

3.5. Transport to the Plasma Membrane

3.5.1. Transport to the Plasma Membrane: Image Acquisition

1. This protocol describes a method which allows the determination of the time required to restore the antagonist-promoted cell surface delivery of an ER-retained GPCR (time dependence). With a similar protocol, it is also possible to analyze the concentrations of antagonists required to restore cell surface delivery maximally (concentration dependence). Following transfection over night, cells are either incubated with a fixed concentration of the antagonist for a variable period (time dependence) or with different concentrations of the antagonist for a fixed period, e.g., 12 h (concentration dependence; *see* **Note 17**).
2. For the analysis of the time-dependence, coverslips are clamped each hour in the holder (*see* **Subheading 2.1.2.**) and are analyzed after addition of 500 μL of buffer. For the analysis of concentration/response effects, cells are analyzed after the end of the chosen incubation period.
3. GFP and trypan blue images should be acquired in the "Multi-track" mode using the "Line scan" for GFP and trypan blue images (as shown in **Table 2**).
4. The parameters (e.g., detector gain, laser intensity) are adjusted for excitation and recording of GFP.
5. In the sample, a representative number of cells ($n = 30$) with comparable level of GFP emission is analyzed and the precise position of the cells (X, Y, Z) is stored with the function "Mark and find" (*see* **Note 18**).
6. Following the identification of cells, use "Mark and find" to start again with the first position.
7. Add 20 μL of trypan blue to the cells of interest and adjust parameters (e.g., detector gain, laser intensity) for excitation and recording of trypan blue.
8. Generate a two-channel image (GFP, trypan blue).
9. Use "Mark and find" to proceed to the next position and generate the next two channel image.
10. Repeat this procedure until two channel images of all marked positions have been documented. In **Fig. 8**, three representative panels of images are shown, demonstrating the differences in the subcellular distribution of the mV_2R.GFP fusion protein (upper panel) following different periods of antagonist treatment. Trypan blue defining the plasma membrane is shown in the lower panel.

3.5.2. Transport to the Plasma Membrane: Quantitative Analysis of the Images

On the basis of the images, it is possible to quantify the relative amount of intracellular and plasma membrane fluorescence, which depends on the concentration of the investigated antagonist and the incubation time. For the quantitative analysis, several computer programs can be used, which include the software supplied with the LSM or other programs, such as the KS400 or Axovision 4.0 (both from Zeiss).

1. The two channel images (obtained in **Subheading 3.5.1.**) are submitted to the analysis.

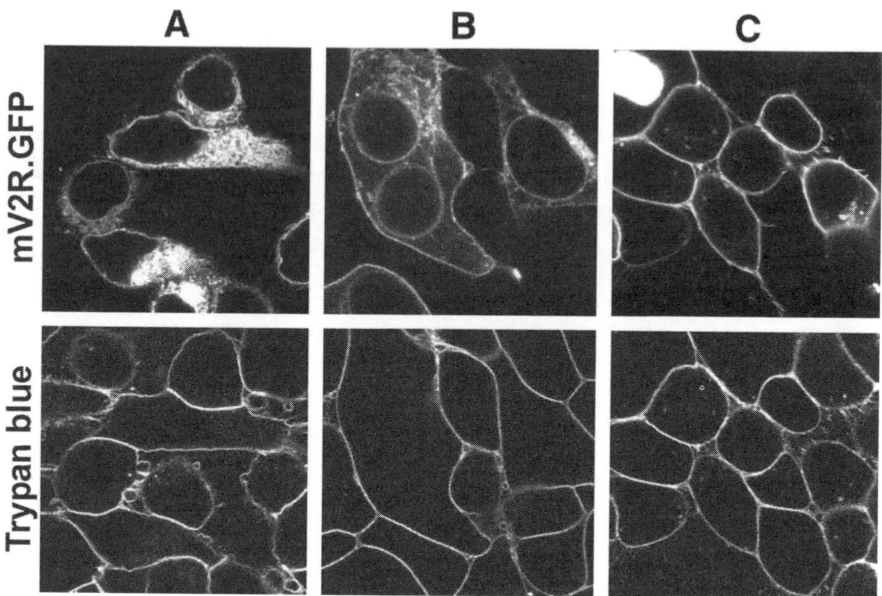

Fig. 8. HEK293 cells transiently expressing the mV2R.GFP were treated with the vasopressin receptor antagonist SR49059 for up to 13 h (**A**: 0 h; B: 7 h; **C**: 13 h). ***Top panel***: mV_2R.GFP fusion protein. ***Bottom panel***: plasma membrane stained with trypan blue. GFP, green fluorescent protein.

2. On the basis of the trypan blue images, the cell membranes are marked as regions of interest (ROIs). The ROIs are transferred to the corresponding GFP image.
3. A second ROI representing the cell's interior is defined in the GFP images.
4. Determine average intensity of the GFP fluorescence for the plasma membrane and the cell's interior.
5. Calculate the ratio (R) of the two intensity values (R = fluorescence of the membrane/cell interior).
6. Generate a graphical presentation. Set the incubation time t (R = f[t]) as the value for the x axis, and the corresponding ratio R as the value for the y axis.
7. A curve of this time-lapse or concentration response analysis can be fitted by a mathematical function, which is $R = R_{min} + D/(1+ e^{[(\tau - t)/A]})$. R_{min} represents the lowest ratio value, D the difference between the highest and lowest ratio value ($R_{max} - R_{min}$), τ the half-life time, and A represents a constant for the increase in the ratio. An example for a time-lapse study is shown in **Fig. 9**. In **Fig. 10**, a concentration-response curve is presented. Such results can be described with an equation for dose response relationship: $R = (R_{max} \times D)/(K_D + D)$. Here, R_{max} represents the maximal ratio, D the concentration of the antagonist and K_D represents the dissociation constant.

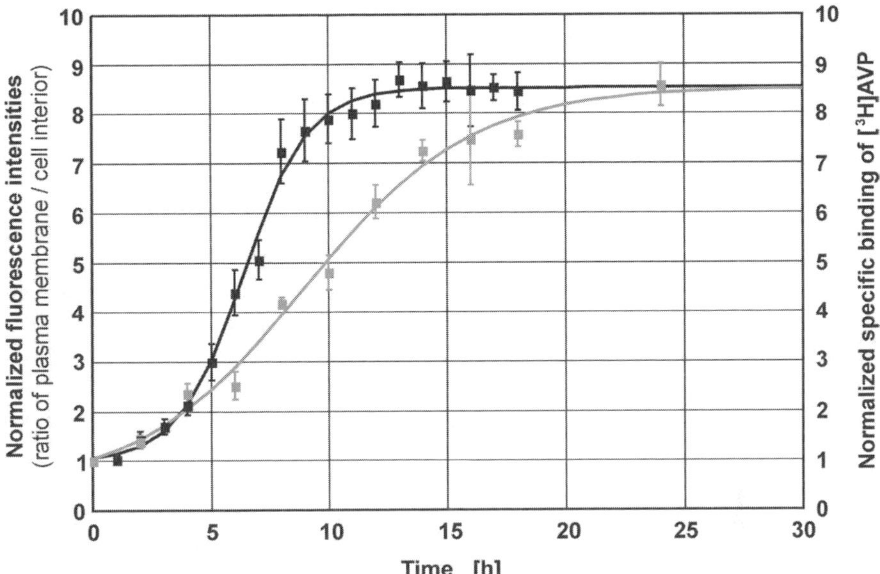

Fig. 9. Antagonist-mediated restoration of cell surface expression analyzed by quantitative laser scanning microscopy (LSM) and binding analysis. Cells transiently expressing the mV_2R.GFP were treated for up 16 h with a specific vasopressin receptor antagonist. Individual samples were analysed each hour by LSM or were subjected each second hour to membrane preparation. Fluorescence of plasma membrane and the cell's interior were quantified and expressed as normalized ratios of the fluorescence intensities (plasma membrane/cell interior). The fitted curve representing the increase in the normalized fluorescence intensities is shown in black. Values are means ± SEM. In parallel, membrane preparations were analyzed for specific binding of $[^3H]$AVP. The curve representing the increase in specifically bound $[^3H]$AVP is shown in grey. Values are means ± SD. GFP, green fluorescent protein.

3.6. Fluorescence Resonance Energy Transfer: Image Acquisition

The settings for FRET analysis with the LSM510 META system are shown in **Table 3**. Images are acquired in the "Multi-track" mode using the "Line mode" for scanning.

1. Insert coverslip with cells expressing a GPCR.CFP fusion protein and cover cells with incubation buffer (1 mL).
2. Adjust laser intensity (458 nm) and the detector gains for all three channels, so that excitation of CFP results in a clear signal in the CFP channel. No signals should be obtained in the YFP and the FRET channel (*see* **Note 19**).
3. In a second step, analyze a coverslip with cells expressing a GPCR.YFP fusion protein. Adjust the laser intensities (458 nm, 514 nm) and the detector gains (all

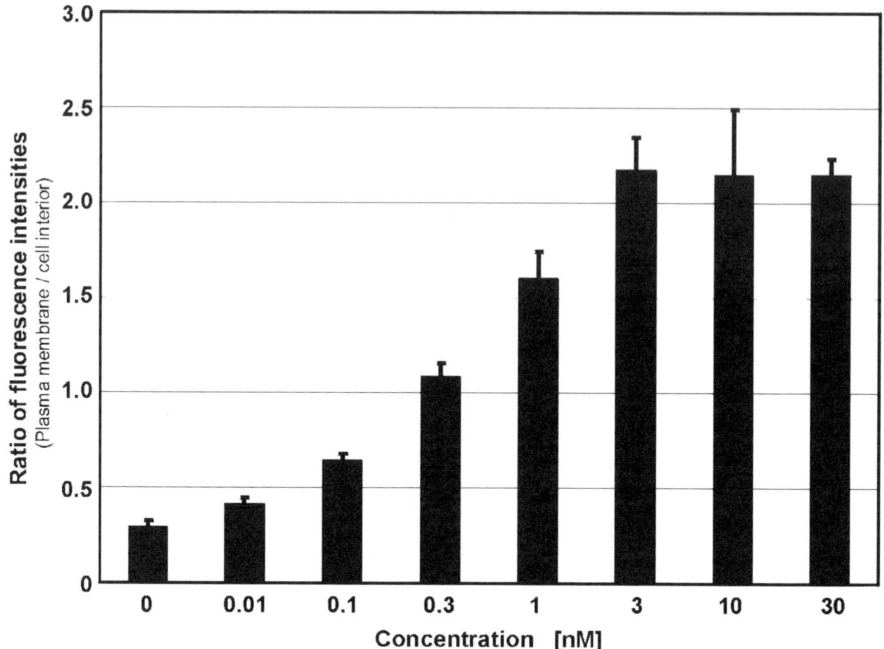

Fig. 10. Concentration response curves of the antagonist-promoted cell surface delivery of the endoplasmic reticulum-retained mV_2R.GFP. HEK293 cells transiently expressing the mV_2R.GFP were treated with a specific vasopressin receptor antagonist in different concentrations for 16 h. Fluorescence of plasma membrane and the cell's interior were quantified and expressed as normalised ratios of fluorescence intensities (plasma membrane/cell interior). Values are means ± SEM. GFP, green fluorescent protein.

three channels), so that excitation of the YFP sample results in a clear signal in the YFP channel, but that no signals are seen in the CFP and the FRET channel.

4. Measure the fluorescence intensity of the cells expressing the GPCR.YFP fusion proteins using the "Histogram" function.
5. Replace the coverslip with a sample in which cells express a GPCR.CFP fusion protein. Determine and note the fluorescence intensity of the CFP sample.
6. Replaced the coverslip with a sample, in which cells co-express GPCR.CFP and GPCR.YFP fusion proteins (*see* **Note 20**). Identify cells for which the intensity values do not exceed the values noted for CFP and YFP. Generate an image for all three channels (CFP, YFP, FRET).
7. Activate the "Time series" and the "Edit bleach" menus.
8. To get a baseline recording for four images without bleaching, followed by a bleaching period with 10 cycles set in the menu "Edit Bleach" the option "Bleach after number scans" to 4 and the option "Bleach repeat after number scans" to 1

Table 3
Laser Scanning Microscopy (LSM) 510 META Settings for Fluorescence Resonance Energy Transfer (FRET) Experiments

Signal	Excitation wavelength	Main beam splitter	META channel	META band-pass
CFP	458 nm	HFT548/514	1	462 nm–500 nm
YFP	514 nm	HFT548/514	2	530 nm–600 nm
FRET	458 nm	HFT548/514	3	530 nm–600 nm

Analysis is performed in the "Multi-track" mode and using the "Line mode" for scanning the samples in the three different channels. CFP, cyan fluorescent protein; YFP, yellow fluorescent protein.

and in the menu "Time series control" the option "Stop series" to 14. For bleaching of certain cells or distinct cellular regions, activate the option "Define ROI" in the menu "Edit Bleach" and set a single or multiple ROIs. With the option "Bleach parameter" the time of bleaching is defined. Activate the laser line for 514 nm and set the "Output" to 100%. Finally define the interval between the different cycles by the option "Cycle delay" in the menu "Time series control." This shortest interval possible is determined by the scanning time for a single image which is indicated in the menu "Scan control".

9. Start the time series, and 4 images without and 10 images with repetitive bleaching will be recorded (*see* **Fig. 5**).
10. The prebleaching and postbleaching images are subjected to further quantitative analysis (*see* **Subheading 3.6.2.**).

3.6.1. Fluorescence Resonance Energy Transfer: Quantitative Analysis

With the quantitative analysis of the stored images (before and after the acceptor bleaching), the FRET efficiency can be calculated ***(29)***. On the basis of these data, it is even possible to show whether FRET efficiencies of GPCRs differ between distinct cellular compartments (e.g., ER and plasma membrane).

1. Reload the three channel images (pre- and postbleaching).
2. Activate the function "Histogram" in the LSM software and mark defined regions as ROI. Determine the mean fluorescence intensities for CFP, YFP, and FRET in the ROIs of the pre- and postbleaching images.
3. The values obtained for CFP, YFP, and FRET emissions before acceptor bleach are set to 100%, and the relative values for CFP, YFP, and FRET after acceptor bleach are calculated. In **Fig. 5**, an example of such a FRET experiment with photobleaching of the acceptor is shown.
4. The efficiency (E) for the FRET can be calculated according to the equation $E = 1 - (F_{DA}/F_D)$. F_{DA} represents the fluorescence intensity of the donor (CFP) in the presence of the acceptor (YFP), and F_D the fluorescence intensity of the donor in the absence of the acceptor.

4. Notes

1. When coverslips are coated with poly-L-lysine, avoid leakage of the solution beneath the coverslip. This spillover is difficult to remove by washing, but can be released into the medium during cultivation and impair proper growth of the cells. Generally, we use only 300 μL of poly-L-lysine solution per 24–30 mm coverslip. Precoated coverslips should be used within a few days. After seeding cells onto coverslips, make sure that coverslips are attached to the bottom of the culture dish. In the case the coverslips are floating on the medium, cells will also grow on the bottom of the coverslip. This can affect the quality of the images.
2. For the analysis of fluorescent proteins, the use of nonconfluent cultures is highly recommended (usually 30–50% of confluence). However, for the analysis of polarized cells or of intercellular interactions (e.g., tight junctions) confluent cell cultures will be required.
   The analysis of polarized cells, such as Madin-Darby canine kidney (MDCK) cells should be not be performed on coverslips, but on semipermeable transwell inserts, which are available with transparent and nontransparent filters from Becton Dickinson or Corning (NY, USA). For LSM, these filters have to be cut out and transferred to a coverslip inserted in a holder (*see* **Subheading 2.1.2.**) with the cells upside down. Place a small coin (e.g., one cent) on the backside of the filter to prevent floating and add 500 mL of buffer. Alternatively, the cells can be fixed on the transwell inserts before being cut out and transferred to a slide (with the cells upside-up). The filter is then mounted with 5 μL Immu-mount (Shandon, Pittsburgh; USA) and sealed with a coverslip.
3. A variety of different transfection methods can be applied (lipofection, electroporation, nucleofection, calcium phosphate precipitation). There is no recommendation for a particular method. However, for LSM each transfection system has be to optimized, irrespective of whether transfection protocols already exist for biochemical assays. High levels of expression can affect the sorting of the investigated proteins, so that the amount of DNA used for transfection experiments has to be determined experimentally. The analysis of stably expressing cell clones is often superior to transiently transfected cells, because stable cell clones usually have a lower level of expression. For co-transfection experiments, the optimal ratio for the transfection of two or more plasmid cDNAs has to be determined experimentally. It is also important to analyze the time required for expression and proper sorting. For many GPCRs, plasma membrane staining can be observed within a few hours after transfection and is optimal after 12–24 h. However, some proteins fold and assemble very inefficiently, so that it may take 48–96 h until proper delivery to the plasma membrane is observed.
4. For the analysis of cells by LSM use phenol-red free medium or buffer, as phenol-red itself can be fluorescent and impair image quality.
5. For internalization experiments, it is most important to keep cells at 37°C. Therefore, a temperable insert has to be used. However, it has to be considered that the temperature of the medium in the proximity of the cells is usually lower than the value set at the thermostat. Therefore, the temperature should be measured in the

medium and the temperature at the thermostat has to be adjusted accordingly. A problem in living cell microscopy is the loss of caloric energy along the objective, which affects, in particular, the temperature around the cells under investigation. Conversely the caloric energy from the heating unit is transferred to the scanning stage, which can affect the correct Z position. This is a serious problem in longer time series experiments as cells move out of focus. These problems can be avoided when a heating/incubation chamber, which surrounds the scanning stage, and the objectives, is used, thereby maintaining sample, scanning stage and objectives at a constant temperature. Such heating/incubation chambers are commercially available for some Zeiss microscopes (PeCon).

6. YFP samples can be analyzed at 488 nm. However, for FRET analysis involving photobleaching of YFP (*see* **Subheading 3.6.**), excitation of YFP at 514 nm is recommended. Excitation of CFP at 458 nm also results in excitation of YFP. Here, excitation of CFP at 405 nm is recommended.
7. Usually, the ligand-induced internalization is a rather slow process which could be monitored in the "frame mode." However, because activation of a variety of GPCRs cause an increase in cytosolic calcium and/or the activation of the small GTPase RhoA, rapid changes in the cellular morphology by alterations in the actomyosin system occur. Thus, it is recommended to analyze the transport of GPCRs in the "line mode."
8. In the "lambda mode," the emission of a fluorescent sample is presented in a gallery of λ-images. A single λ-image represents a spectral window of only 10 nm. As a consequence, the single λ-image will be of a low signal intensity. Following linear unmixing, the images will yield a stronger signal intensity, as the intensities from the different λ-images are superimposed. A common mistake is that the laser intensity is increased to obtain stronger signals in a single λ-image. However, the increased excitation results in significant bleaching of the fluorophores.

   For "emission fingerprinting," the spectral signature of each fluorophore is required. These spectra have to be acquired and stored in the dye data base of the LSM510 META system before starting the mode "linear unmixing." It is of note that the quality of the linear unmixing depends completely on the accuracy of the spectral signature of each fluorophore. As the spectral signatures of fluorophores can vary depending on the environment (pH, concentration of cations, redox potential), the spectral signature of each fluorophore should be determined in a defined cellular compartment. In addition, it is also of relevance that the laser lines, beam splitters, and filters are identical to those used for the determination of the spectral signature of a single fluorophore.
9. Labeling of the plasma membranes with trypan blue is a fast process. However, the diffusion rate of trypan blue when applied to the medium is very slow. Therefore, 20 μL of trypan blue should be directly administered to the cells under investigation. Because trypan blue does not enter living but dead cells, it is also a good marker for the cell viability. Most of the live cell markers (e.g., trypan blue, the bisbenzimide dye H33258, rhodamine 6G, ceramide) need not be removed

after addition to the medium. The markers either reveal fluorescence only when bound to the plasma membrane or the organelle (trypan blue, H33258), or the markers are only seen following accumulation within an organelle (rhodamine 6G, ceramide). A variety of chemical markers also bind unspecifically to the coverslip holder and are released during the analysis of further samples. Especially in the case of rhodamine 6G the coverslip holder has to be extensively washed. For rhodamine 6G, washing with ethanol is recommended.

10. The labeling of peptides on a small scale (less than 300 μg) results in poor recoveries during the purification process. The total amount of labeled ligand cannot be determined accurately. However, for the functional testing the molarity of the labeled peptide is needed.
11. The Cy- and Alexa-NHS esters are supplied in a dried, premeasured formulation and can be used immediately. If only a fraction of the reactive Cy-NHS ester is required, dissolve the dyes in anhydrous DMSO and make aliquots (store in the freezer in light-protected tubes).
    It is recommended to verify the quality of NHS esters by analytical HPLC before starting the labeling reaction. When a significant amount of the NHS ester is already hydrolyzed, no—or only insufficient—dye incorporation is achieved.
    Because the different commercially available dyes differ in their physical and chemical properties (pH sensitivity, hydrophobicity, charge, molecular weight) there is no general recommendation for a certain dye. Loss of functional activity may be observed with some dyes, but may be preserved with others. The ideal dye has to be determined experimentally. If possible, avoid dyes for which isomers exist, e.g., where the position of the reactive carboxyl group added to the aromatic group in the dye differs within the preparation. Such isomers present a problem in analytical and functional testing.
    Buffers with free amino groups, such as Tris-HCl or glycine, may reduce the labeling efficiency of peptide ligands. However, in our hands, labeling of the N-terminal amino group in Tris buffer did not reduce labeling efficiency.
12. Verification of dye incorporation and NHS ester consumption by analytical HPLC (equipped with a UV and fluorescence detector) is important. When NHS ester is still present in the reaction, increase the pH and continue the reaction. When dye was insufficiently incorporated into the peptide, the pH value of the reaction may be to high. Control and adjust pH value and supply reaction with additional NHS ester.
13. For functional testing of fluorescent ligands, binding analysis is recommended as the first step. In displacement-binding experiments, changes in the affinities caused by the fluorophore can be determined. For these experiments, only small amounts of the ligand are required. After defining the affinity of the fluorescent ligand, the agonistic activity should be verified (for example, calcium imaging, GTPγS binding, inositol phosphate production).
14. Use only small incubation volumes for the analysis of cells to keep the amount of fluorescent ligands as small as possible. It is highly recommended that the fluorescent ligand be added in a volume equalling the volume of the medium in the

coverslip holder (here 250 µL are used). Thereby, an almost complete mixing of both buffers is guaranteed and ligand–receptor interactions will not be limited by diffusion.

When the coverslip is mounted in the temperable insert, beware of evaporation. Because the cells are covered with only 250 µL of medium, the sample may dry out during adjustment of the parameters. Evaporation can be reduced when the coverslip holder is covered with a lid of a Petri dish and when LSM parameters for the analysis are adjusted quickly. As the fluorescent ligand is added to the cells during the time series, make sure that the lid can be quickly removed. Otherwise, remove the lid when starting the time series.

15. In the time series, the interval between single scans should be chosen according to the experimental purpose. When analyzing internalization of a GPCR, a period of approx 10–60 min might be analyzed. As prolonged scanning of the cells causes bleaching of the fluorophore and can be harmful for the cells (e.g., lipid peroxidation), longer intervals (1–5 min) are recommended. When the association of the fluorescent ligand with the receptor is analyzed, for most peptide ligands the association is complete within 5–10 min. Here, intervals of 10 to 30 s are recommended.
16. Receptor–ligand association can be quantified by the menu "histogram." Here, ROIs in which the intensities of fluorescence over the complete time series is depicted can be defined. For GPCRs with a single binding site, a simple saturation curve will be obtained, which allows determination of the half maximal time required for receptor binding.
17. This protocol is for the analysis of misfolded, ER-retained GPCRs, which show antagonist-mediated restoration of cell surface expression. Such an ER retention, which can be overcome by antagonist- or inverse agonist-treatment, has been observed for several mutant GPCRs, including rhodopsin and the gonadotropin-releasing hormone receptor, as well as wild-type GPCRs such as the δ-opioid receptor and the murine V_2R ***(23–26,30)***. However, it is possible to study the ER to plasma membrane transport with any other GPCR using drugs which result in the accumulation of GPCRs in the ER. ER accumulation can be induced by pretreatment with brefeldin A (10 µ*M*). Brefeldin A can be washed out and transport of the GPCR from the ER to the plasma membrane can be monitored.
18. The "Mark and find" option can only be used when the LSM510 META system is equipped with a motorised stage. With a manual stage, the XYZ positions have to be noted and cells repositioned manually.
19. FRET experiments with CFP and YFP fusion proteins using the LSM, several problems have to be considered. When CFP is excited at 458 nm, this will also lead to an excitation of YFP. Excitation of CFP with a violet diode laser at a wavelength of 405 nm or by two-photon absorption at 820 nm will improve excitation of CFP and, in parallel, reduce excitation of YFP.

    A further problem is that the long-wave emission of the donor (CFP) is also detected in the acceptor (YFP) channel. Here, FRET has to be proved by photobleaching of the acceptor (YFP). In the case of a protein–protein interaction, the destruction of YFP will result in an increase in the emission of CFP. It is notable that also the donor (CFP) shows destruction, especially when the sample is bleached at 488 nm and not at 514 nm as recommended.

20. When starting with FRET analysis, the ratio of CFP to YFP fluorescence has to be determined experimentally, as the FRET efficiency varies with the formation of dimers or higher order oligomers. In most cases, a two- to threefold excess of the acceptor over the donor yields the best results.
    Although CFP and YFP have only a weak tendency to dimerize, it might be possible that upon high levels of expression or clustering of fusion proteins in a cellular compartment, the dimerization of CFP / YFP moieties can result. To exclude such a GFP-mediated dimerization, FRET should be verified by competition experiments. Here, cells are co-transfected with plasmids encoding the GPCR.CFP and GPCR.YFP fusion proteins and, in addition, a plasmid encoding the native, non-CFP/YFP-tagged GPCR. When FRET is caused by direct GPCR interactions, the FRET efficiency will decrease by the co-expression of native GPCR. In the case of GFP-mediated dimerization, no effect of co-transfection will be observed. Alternatively, GFP-mediated dimerization can be excluded by the use of monomeric CFP/ GFP/YFP proteins. The amino acids involved in the formation of GFP-dimers are A206, L221, and F223. The substitution of alanine 206 by lysine (A206K) is sufficient to eliminate the dimerization of CFP/GFP/YFP proteins (***4***).
    For FRET experiments, negative and positive controls should be included. A negative control may be represented by the co-expression of CFP and YFP only. As a positive control, a CFP.YFP tandem protein can be used ***(31)***.

Acknowledgments

We thank Jenny Eichhorst for excellent technical assistance, Brunhilde Oczko for assistance in analyzing of the microscopic data, Tim Plant for reading the manuscript and providing helpful comments, and Jürgen Mevert and Stephanie Wendt for manufacturing customized cuvettes for the LSM experiments.

References

1. Prasher, D.C., Eckenrode, V.K., Ward, W.W., Prendergast, F.G., and Cormier, M.J. (1992) Primary structure of the *Aequorea victoria* green-fluorescent protein. *Gene* **111**, 229–233.
2. Prasher, D.C. (1995) Using GFP to see the light. *Trends Genet.* **11**, 320–323.
3. Tsien, R.Y. (1998) The green fluorescent protein. *Annu. Rev. Biochem.* **67**, 509–544.
4. Zhang, J., Campbell, R.E., Ting, A.Y., and Tsien, R.Y. (2002) Creating new fluorescent probes for cell biology. *Nat. Rev. Mol. Cell. Biol.* **3**, 906–918.
5. Ellenberg, J., Lippincott-Schwartz, J., Presley, J.F. (1999) Dual-colour imaging with GFP variants. *Trends Cell Biol.* **9**, 52–56.
6. Kallal, L., and Benovic, J.L. (2000) Using green fluorescent proteins to study G protein-coupled receptor localization and trafficking. *Trends Pharmacol. Sci.* **21,** 175–180.

7. Milligan, G. (1999) Exploring the dynamics of regulation of G protein-coupled receptors using green fluorescent protein. *Br. J. Pharmacol.* **128**, 501–510.
8. Milligan, G. (2000) Insights into ligand pharmacology using receptor-G-protein fusion proteins. *Trends Pharmacol. Sci.* **21**, 24–28.
9. Azpiazu, I. and Gautam, N. (2004) A FRET based sensor indicates that receptor access to a G-protein is unrestricted in a living mammalian cell. *J. Biol. Chem.* **279,** 27,709–27,718.
10. Overton, M.C. and Blumer, K.J. (2002) Use of fluorescence resonance energy transfer to analyze oligomerization of G protein-coupled receptors expressed in yeast. *Methods* **27**, 324–332.
11. Gregan, B., Jürgensen, J., Papsdorf, G., et al. (2004) Ligand-dependent differences in the internalization of endothelin A and endothelin B receptor heterodimers. *J. Biol. Chem.* **279,** 27,679–27,687.
12. Baird, G.S., Zacharias, D.A., and Tsien, R.Y. (2000) Biochemistry, mutagenesis, and oligomerization of DsRed, a red fluorescent protein from coral. *Proc. Natl. Acad. Sci. USA* **97**, 11,984–11,989.
13. Daly, C.J. and McGrath, J.C. (2003) Fluorescent ligands, antibodies, and proteins for the study of receptors. *Pharmacol. Ther.* **100**, 101–118.
14. Oksche, A., Boese, G., Horstmeyer, A., et al. (2000) Late endosomal/lysosomal targeting and lack of recycling of the ligand-occupied endothelin B (ET_B) receptor. *Mol. Pharmacol.* **57**, 1104–1113.
15. Ferguson, S.S. (2001) Evolving concepts in G protein-coupled receptor endocytosis: the role in receptor desensitization and signaling. *Pharmacol. Rev.* **53**, 1–24.
16. Ferguson, S.S., Zhang, J., Barak, L.S. and Caron, M.G. (1998) Molecular mechanisms of G protein-coupled receptor desensitization and resensitization. *Life Sci.* **62**, 1561–1565.
17. Gurevich, V.V. and Gurevich, E.V. (2004) The molecular acrobatics of arrestin activation. *Trends Pharmacol. Sci.* **25**, 105–111.
18. Ferguson, S.S., and Caron, M.G. (2004) Green fluorescent protein-tagged β-arrestin translocation as a measure of G protein-coupled receptor activation. *Methods Mol. Biol.* **237**, 121–126.
19. Luttrell, L.M. and Lefkowitz, R.J. (2002) The role of β-arrestins in the termination and transduction of G protein-coupled receptor signals. *J. Cell Sci.* **115**, 455–465.
20. Mousavi, S. A., Malerod, L., Berg, T. and Kjeken, R. (2004) Clathrin-dependent endocytosis. *Biochem. J.* **377**, 1–16.
21. Schulz, R., Wehmeyer, A., and Schulz, K. (2002) Opioid receptor types selectively cointernalize with G protein-coupled receptor kinases 2 and 3. *J. Pharmacol. Exp. Ther.* **300**, 376–384.
22. Barak, L.S., Wilbanks, A.M., and Caron, M.G. (2003) Constitutive desensitization: a new paradigm for G protein-coupled receptor regulation. *Assay Drug Dev. Technol.* **1**, 339–346.
23. Janovick, J.A., Maya-Nunez, G., and Conn, P.M. (2002) Rescue of hypogonadotropic hypogonadism-causing and manufactured GnRH receptor mutants by a specific protein-folding template: misrouted proteins as a novel disease etiology and therapeutic target. *J. Clin. Endocrinol. Metab.* **87**, 3255–3262.

24. Morello, J.P., Salahpour, A., Laperriere, A., et al. (2000) Pharmacological chaperones rescue cell-surface expression and function of misfolded V_2 vasopressin receptor mutants. *J. Clin. Invest.* **105**, 887–895.
25. Noorwez, S.M., Kuksa, V., Imanishi, Y., et al. (2003) Pharmacological chaperone-mediated in vivo folding and stabilization of the P23H-opsin mutant associated with autosomal dominant retinitis pigmentosa. *J. Biol. Chem.* **278**, 14,442–14,450.
26. Wüller, S., Wiesner, B., Löffler, A., et al. Pharmacochaperones posttranslationally enhance cell surface expression by increasing conformational stability of wild-type and mutant vasopressin V_2 receptors. *J. Biol. Chem.* **279,** 47,254–47,263.
27. Vilardaga, J.-P., Bünemann, M., Krasel, C., Castro, M. and Lohse, J.L. (2003) Measurement of the millisecond activation switch of G protein-coupled receptors in living cells. *Nature Biotech.* **21**, 807–812.
28. Damke, H., Baba, T., Warnock, D.E., and Schmid, S.L. (1994) Induction of mutant dynamin specifically blocks endocytic coated vesicle formation. *J. Cell Biol.* **127**, 915–934.
29. Gordon, G.W., Berry, G., Liang, X.H., Levine, B., and Herman, B. (1998) Quantitative fluorescence resonance energy transfer measurements using fluorescence microscopy. *Biophys. J.* **74**, 2702–2713.
30. Petaja-Repo, U.E., Hogue, M., Bhalla, S., Laperriere, A., Morello, J.P. and Bouvier, M. (2002) Ligands act as pharmacological chaperones and increase the efficiency of delta opioid receptor maturation. *EMBO J.* **21**, 1628–1637.
31. Lenz, J.C., Reusch, H.P., Albrecht, N., Schultz, G., and Schaefer, M. (2002) Ca2+-controlled competitive diacylglycerol binding of protein kinase C isoenzymes in living cells. *J. Cell Biol.* **159**, 291–302.

10

Imaging and Characterization of Radioligands for Positron Emission Tomography Using Quantitative Phosphor Imaging Autoradiography

Peter Johnström and Anthony P. Davenport

1. Introduction

1.1. Phosphor Imaging

Quantification of radioligand binding by exposing labeled tissue sections to phosphor screens in cassettes is similar to autoradiography using radiation sensitive film *(1)* (*see* Chapters 5 and 7). The major advantage of phosphor screens over film is the greatly increased sensitivity with exposure times reduced by at least one order of magnitude *(2,3)*. This is essential for short-lived radionuclides—such as ^{18}F and ^{11}C, with half-lives of 109.8 and 20.4 min, respectively—that are used to label peptide or drug ligands to image receptors noninvasively in vivo by positron emission tomography (PET) (*see* Chapter 11). Phosphor imaging can also considerably reduce exposure times for weak β-particle emitters such as ^{3}H from months to days *(3)*. A second advantage is the increased linear dynamic range of five orders of magnitude *(2,3)*. This increased dynamic range makes it less likely that screens will be saturated. This is important for PET radioligands in which, owing to the short half-life of the radionuclide, it is only possible to image once, whereas re-exposure is possible for isotopes with longer half-lives.

A limitation with phosphor imaging compared with film can be the decreased resolution, which is most noticeable for low-energy isotopes such as ^{3}H and ^{125}I. However, for macroautoradiography applications using ^{14}C, spatial resolution comparable with that of film is obtained *(2)*. Consequently, for the high-

From: *Methods in Molecular Biology, vol. 306: Receptor Binding Techniques: Second Edition*
Edited by: A. P. Davenport © Humana Press Inc., Totowa, NJ

energy radionuclides used in PET, this limitation will not be of any significance.

In this technique, radioactivity is detected by a thin layer of BaFBr:Eu^{2+} phosphor crystals (instead of silver halide as in film autoradiography) usually protected by moisture-proof coating. Samples containing radioactivity are apposed to the screen, which can be gel, filter paper, capillary tube, or—as described in detail in this chapter—sections of tissue mounted on microscope slides.

The phosphor crystals absorb and store the energy of the radioactive emission. In this process, Eu^{2+} is oxidized to Eu^{3+} and the released electron is stored in the phosphor lattice. At the end of the exposure period, the screen is scanned in a phosphor imager using a red laser, which will release the trapped electron, converting Eu^{3+} back to Eu^{2+} and re-emitting the stored energy as blue light. The intensity of the emitted light is proportional to the amount of radioactivity in the sample. The blue light is detected by a photomultiplier tube and the data are stored as a digital image of the locations and intensities of the radioactivity in the sample. The produced image (autoradiogram) can then be analyzed using image analysis software and, if a set of standards are included with the samples, the amount of radioactivity within discrete regions of the image can be quantified. The screen can be re-used simply by erasing the stored images with white light from a light box (*see* **Note 1**).

Phosphor imaging systems are available commercially from Fuji Medical Systems (http://www.fujimed.com) and Amersham Biosciences (http://www4.amershambiosciences.com), and the system used in this chapter, the Cyclone Phosphor Imager, is available from PerkinElmer Life Sciences (http://las.perkinelmer.com).

1.2. Application of Phosphor Imaging

PET is the only technique to image and quantify the amount of receptor-bound ligand in vivo. A key step in the synthesis of a novel radioligand is to ensure that the desirable binding properties of the unlabeled compound (high affinity and selective and reversible binding) have been maintained despite the introduction of the radionuclide into the molecular structure. This can be assessed by characterizing the behavior of the radioligand using in vitro binding assays and phosphor imaging autoradiograhy.

Furthermore, phosphor imaging is a powerful tool for rapid ex vivo analysis of whole-body or tissue sections from rat or mouse to visualize distribution of infused short-lived PET radioligands in animal experiments. In combination with small animal PET imaging using tomographs such as the microPET (*see* Chapter 11), ex vivo autoradiography will visualize the distribution of radioligand to a higher resolution, particularly on the suborgan level, than that obtained in vivo with PET, verifying that radioligand uptake is localized to areas with receptor densities. To further validate a novel radioligand, tissue section can be left for an appropriate time period to allow for the short-lived radioactivity to decay, and subsequently be incubated in vitro with a radioligand well-characterized for the receptor system studied. Tissue section should be stored at –70°C to ensure that receptor proteins are not degraded. If the anatomical distribution obtained in vitro, reflecting specific binding to receptors, matches that obtained in vivo/ex vivo this is further proof that the PET radioligands have the potential to image the receptor system in vivo with PET.

Phosphor imaging can be used in dual isotope in vivo/ex vivo applications using a high-energy, short-lived PET radionuclide (e.g., ^{18}F) in combination with a low-energy radionuclide (e.g., ^{14}C). Exposing the screen twice to the dual-labeled tissue, in combination with a time period between exposures to allow for the decay of the short-lived PET radionuclide, will yield separate images reflecting the distribution of the two radioligands. For example, the interrelationship between local cerebral blood flow (LCBF) and glucose metabolism (LCMRglc) in the same rat after controlled cortical impact injury has been investigated using ^{14}C-iodoantipyrine (^{14}C-IAP) and ^{18}F-fluorodeoxyglucose ([^{18}F]FDG) (***4***). Initially, sections of the dual-labeled brain tissue were exposed to a phosphor screen for 3 h to yield an image predominantly reflecting ^{18}F uptake, because contribution of ^{14}C activity at this exposure time was negligible (<0.1%). The same sections were then re-exposed to the phosphor screen 2 d later, when all ^{18}F had decayed, for 3 d to obtain an image reflecting the ^{14}C uptake and consequently enabling comparison of LCBF and LCMRglc data for the same anatomical region.

2. Materials

2.1. Ligand-Binding Assays

1. Slide-mounted cryostat sections (10–30 µm) from fresh-frozen tissue (*see* Chapter 5).
2. [^{18}F]-labeled radioligands, such as [^{18}F]-ET-1 (***5***), unlabeled ligand to define nonspecific binding (NSB), and competing ligands for specificity assays.
3. Assay buffer. The optimum binding assay should be determined empirically, but a simple buffer for [^{18}F]-ET-1 that may be suitable for other ligands is 0.05 *M* HEPES, 5 m*M* $MgCl_2$, 0.3% bovine serum albumin (BSA) Fraction V, pH 7.4 (*see* **Note 2**). To separate bound from free ligand, sections were washed using ice-cold 0.05 *M* Tris-HCl buffer, pH 7.4, at 4°C.
4. Slide incubation trays with lids, metal slide racks, and slide baths.

2.2. Ex Vivo Analysis of Tissue Sections

1. Single-edged razor blades.
2. Cryostat chucks.

3. Mounting medium, OCT compound Gurr® (361603E Merck/BDH, Poole, Dorset, UK).
4. Cryostat with motorized microtome, to cut frozen sections of 10–30 μm (e.g., Bright Instruments Co. Ltd., Huntingdon, Cambs., UK).
5. Gelatin-coated microscope slides (*see* Chapter 5).

2.3. Materials for Quantitative Phosphor Imaging Autoradiography and Image Analysis

1. Phosphor imaging system (e.g., the Cyclone Phosphor Imager) equipped with software for data acquisition and image analysis (e.g., OptiQuant, PerkinElmer Life Sciences, http://las.perkinelmer.com).
2. Phosphor imaging screen appropriate for the isotope (*see* **Tables 1** and **2** and **Note 3**).

3. Methods

3.1. Saturation-Binding Assays for Characterizing ^{18}F-Labeled Ligands

Receptor affinity and density can be readily determined by saturation analysis. For phosphor imaging, fresh frozen tissue sections (usually 10–30 μm thick) are cut on a microtome and mounted on microscope slides. Saturation assays will be illustrated using [^{18}F]-ET-1, a 21 amino acid peptide *(5)*. For [^{18}F]-ET-1, 90-min incubation at 23°C is sufficient to reach equilibrium *(6)* (*see* **Note 4** and **Subheading 3.3.**). Equilibrium is rapidly broken by washing to separate bound from free ligand, and conditions for this should be examined to find an optimal compromise between retention of bound label and a high percentage of specific binding. Sections are apposed to a phosphor imaging screen to determine the amount of radioactivity bound to the tissue. An example of a saturation isotherm for [^{18}F]-ET-1 to human left ventricle obtained using phosphor imaging is shown in **Fig. 1**.

1. Cut consecutive cryostat sections (typically 10–30 μm) of fresh-frozen tissue and thaw-mount onto gelatin-coated microscope slides. Allow to air-dry briefly, and store at –70°C until required. Typically, 20 sections (10 Total and 10 NSB) are required for each saturation curve together, with a further three sections collected into microcentrifuge tubes to measure protein.
2. Dilute the stock solution of [^{18}F]-ET-1 in assay buffer to give a highest concentration of about 1–2 n*M*.
3. Using the highest concentration, prepare a serial dilution (500 μL label + 500 μL assay buffer) to give a total of 10 concentrations covering the p*M*–n*M* range. Vortex and use a new pipet tip between each dilution.
4. Remove 250 μL from each of the serial dilutions to determine the NSB, leaving 250 μL to measure total binding (Totals). Add 2.5 μL of 0.1 m*M* unlabeled

Table 1
Properties of Phosphor Screens Available for the Cyclone Phosphor Imager (PerkinElmer Life Sciences; ≠http://las.perkinelmer.com)

Screen	Description	Sensitivity[a]	Spatial resolution[b]
MultiPurpose (MP)	General purpose screen. Resistant to moisture.	0.47	0.25
Super Resolution (SR)	Fine grain phosphor crystal layer for high resolution applications. Relatively resistant to moisture.	0.22	0.33
Super Sensitive (ST)	Thick phosphor crystal layer for samples with low level of radioactivity. Coated with thin protective layer, not as moisture proof as MP and SR screens.	1.00	0.17
Tritium Sensitive (TR)	Noncoated screens for the detection of low-energy β-emission from ^3H. Very sensitive to moisture, only recommended for ^3H applications.		0.38

[a]The net response to ^{32}P dots for MP, SR, and ST screens. Values normalized to the screen with the highest response (ST). Data from Upham and Englert *(3)*.

[b]The spatial resolution was measured using the contrast transfer function (CTF) for 2.5 line pairs/mm and ^{14}C. The higher the CTF value, the better the resolution. Data from Upham and Englert *(3)*.

Table 2
Available Sizes of Phosphor Screens for the Cyclone Phosphor Imager (PerkinElmer Life Sciences; http://las.perkinelmer.com)

Screens	Available screen type	Size (cm)	Number of microscope slides imaged at one time[a]
Small	MP, SR, ST, TR	12.5 × 19.2	14
Medium	MP, SR, ST, TR	12.5 × 25.2	18
Large	MP, SR, ST	12.5 × 43.0	32

[a]Microscope slides (7.5 × 2.6 cm) to be arranged in two rows with tissue ends facing each other and frosted ends facing out. Using this arrangement will ensure that all tissue will be imaged, because the width of the screen does not fully cover the length of two microscope slides. For quantitative data, allow for at least one slide to be designated for the standard curve.

ET-1 to each NSB tube to give a final concentration of 1 μ*M* in order to define the NSB.

5. Pre-incubate 20 microscope slides bearing consecutive tissue sections (10 Total, 10 NSB) with 200 μL assay buffer for 15 min at room temperature to remove endogenous ligand and degradative enzymes.
6. Tip off pre-incubation buffer into tray and replace with 200 μL of each Total or NSB solutions. Cover with a lid to maintain the humidity, and incubate for 90 min at room temperature to reach equilibrium.
7. Break equilibrium by transferring slides to racks and washing in 400-mL baths containing ice-cold 0.05 *M* Tris-HCl buffer, pH 7.4, at 4°C (three times for 5 min).
8. Rinse sections once in de-ionized water to remove buffer salts and dry rapidly in a stream of cold air prior to apposing to phosphor imaging screen (*see* **Subheading 3.5.**).

3.2. Competition-Binding Experiments

In competition-binding assays, the ability of unlabeled ligands to compete for the binding of a fixed concentration of the radioligand is tested over a wide concentration range, typically 10 p*M*–100 μ*M* (*see* Chapter 5). This type of experiments can also potentially be performed using PET radioligands *(7)*. However, for novel PET radioligands, competition experiments, are more commonly used to validate that the binding specificity of the radioligand to the receptor has been retained. For these experiments, a single concentration of an unlabeled ligand (with high affinity and specificity to the receptor) is used to block the binding of the PET radioligand to the tissue. Specificity evaluation will be illustrated by the blockade of [^{18}F]-ET-1 binding to ET receptors in human kidney tissue using the ET_A-selective antagonist FR139317 and the

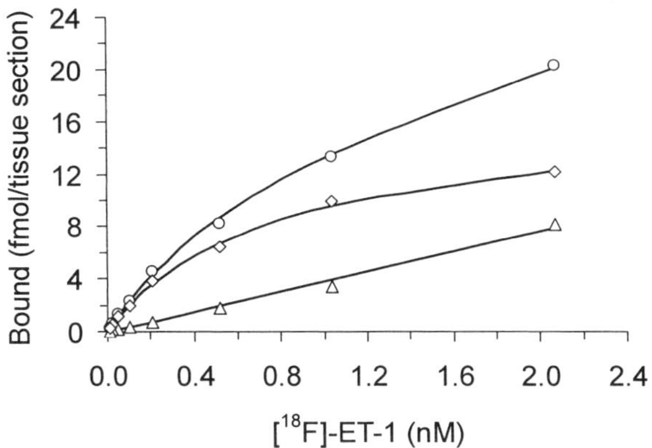

Fig. 1. Representative graphs showing Total (○), Specific (◇), and Nonspecific (△) binding of [^{18}F]-ET-1 to human heart tissue. Nonlinear co-analysis of data from four experiments using the KELL software package showed that binding was concentration-dependent and saturable with a $K_D = 0.43 \pm 0.05$ n*M* and a $B_{max} = 27.8 \pm 2.1$ fmol/mg protein *(6)*. A one-site model was preferred to a two-site model, and the Hill slope (nH) was 0.95 ± 0.04.

ET_B-selective agonist BQ3020 ***(8)***. Both receptor subtypes are present in human kidney, with the ET_B receptor being the predominant subtype (ET_A:ET_B 35%:65%) ***(9)***. **Figure 2** shows phosphor images of a representative blocking experiment. Quantification of phosphor imaging data showed a significant reduction ($p < 0.05$) in specific binding of 34% and 73% when [^{18}F]-ET-1 was co-incubated with FR139317 (ET_A) and BQ3020 (ET_B), respectively, verifying that the radioligand had retained its subtype specificity ***(6)***.

1. Cut four consecutive cryostat sections (typically 10–30 μm) of fresh-frozen tissue per assay, including one section to measure total binding, one section for the NSB, and one section each for investigation of inhibition of binding using the unlabeled ligands FR139317 and BQ3020, respectively. Thaw-mount tissue onto gelatin-coated microscope slides. Allow to air-dry briefly, and store at –70°C until required.
2. Dilute the stock solution of radiolabeled [^{18}F]-ET-1, aiming for a concentration to occupy 10% of the receptors (*see* Chapter 5) in 2 mL of assay buffer. Divide in four test tubes (250 μL each) to define Total binding, add unlabeled ET-1 to a final concentration of 1 μ*M* to define the NSB, and add unlabeled FR139317 (1 μ*M*) and BQ3020 (1 μ*M*) to evaluate specificity for ET_A and ET_B receptors, respectively. Vortex.
3. Pre-incubate four microscope slides bearing consecutive sections with 200 μL assay buffer for 15 min at room temperature.
4. Tip off pre-incubation buffer into tray and replace with 200 μL of the total, NSB, ET_A, and ET_B receptor blockade solutions. Cover with a lid to maintain the humidity and incubate for 90 min at room temperature to reach equilibrium.

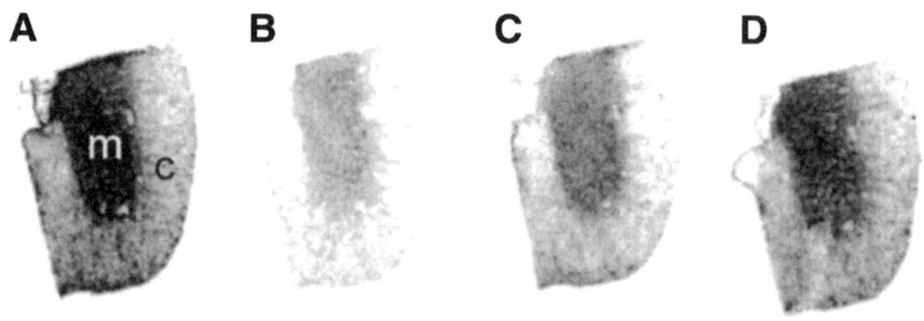

Fig. 2. Phosphor images showing the distribution of [^{18}F]-ET-1 in adjacent sections of human kidney and blockade of binding using ET_A- and ET_B-selective ligands: image showing total binding with the expected high densities of ET receptors in medulla (m) and low densities in cortex (c) (**A**); image showing nonspecific binding (NSB). NSB was defined by co-incubation with unlabeled ET-1 (1 μ*M*) (**B**); image showing partial inhibition of [^{18}F]-ET-1 binding when the tissue was co-incubated with the ET_B-selective agonist BQ3020 (1 μ*M*) (**C**), and with the ET_A-selective antagonist FR139317 (1 μ*M*) (**D**).

5. Break equilibrium by transferring slides to racks and washing in 400-mL baths containing ice-cold 0.05 *M* Tris-HCl buffer, pH 7.4, at 4°C (three times for 5 min).
6. Rinse sections once in de-ionized water to remove buffer salts and dry rapidly in a stream of cold air prior to apposing to phosphor imaging screen (*see* **Subheading 3.5.**).

3.3. Binding Kinetics

Kinetic experiments determine the time course of ligand association and dissociation as described in Chapter 5. For association experiments, sections are incubated with a fixed concentration of radioligand for increasing time periods until equilibrium is reached. When equilibrium is reached, dissociation of the radioligand from the receptors is achieved either by incubation of tissue sections with a high concentration of unlabeled competitor or by infinite dilution of the labeled sections by immersion in a large volume of buffer.

Kinetic assays can be used to validate the potential of a novel radioligand to image receptors in vivo using PET. Ideally, a PET radioligand should have a fast association rate to ensure that binding to the receptors will take place within the time available for a PET scan. The data obtained in an association experiment are also essential when performing saturation (*see* **Subheading 3.1.**) and competition assays (*see* **Subheading 3.2.**) to ensure that measurements are performed when ligand binding has reached equilibrium. Association experi-

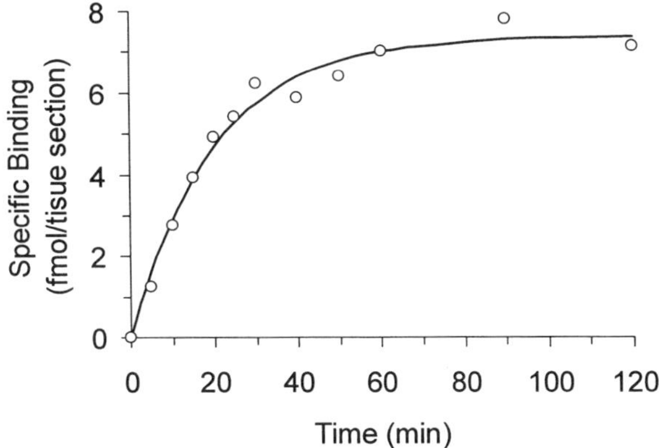

Fig. 3. Association curve for a fixed concentration of [^{18}F]-ET-1 binding to sections of human left ventricle. The curve plateaus after 90 min, indicating that [^{18}F]-ET-1 binding has reach equilibrium. The calculated association rate constant (k_{obs}) was 0.045/min (***6***).

ments will be illustrated using [^{18}F]-ET-1 *(5)*. A representative association curve for [^{18}F]-ET-1 is shown in **Fig. 3**.

1. Cut 20 consecutive cryostat sections (typically 10–30 µm) of fresh-frozen tissue per association assay. One section is used to measure total binding, and one section for the NSB at each of 10 time points. Thaw-mount tissue onto gelatin-coated microscope slides. Allow to air-dry briefly and store at –70°C until required.
2. Dilute the stock solution of radiolabeled [^{18}F]-ET-1, aiming for a concentration to occupy 10% of the receptors (*see* Chapter 5) in 5 mL of assay buffer. Remove 2.5 mL to define Total binding, and add unlabeled peptide to a final concentration of 1 µ*M*, to define the NSB. Vortex.
3. Pre-incubate 20 microscope slides bearing consecutive sections with 200 µL assay buffer for 15 min at room temperature.
4. Tip off pre-incubation buffer into tray and replace with 200 µL of Total (10 slides) or NSB (10 slides). Transfer slides at intervals of 5, 10, 15, 20, 25, 30, 40, 50, 60, 90 and 120 min into racks, and wash in 400-mL baths containing ice-cold 0.05 *M* Tris-HCl buffer, pH 7.4, at 4°C (three times for 5 min).
5. Rinse sections once in de-ionized water to remove buffer salts and dry rapidly in a stream of cold air prior to apposing to phosphor imaging screen (*see* **Subheading 3.5.**).

3.4. Ex Vivo Imaging of Tissue Sections From Animal Experiments

All animal experiments must be conducted in accordance with the appropriate legislation for experimental work on animals and comply with guidelines of the local animal ethics committee. In the United Kingdom, experimental work on animals is regulated by the UK Animal Scientific Procedures Act, 1986. At the end of the in vivo experiment, the animal is euthanized using an approved method, and organs of interest are collected for cryostat sectioning. When working with short-lived PET radionuclides, tissue has to be cryostat-cut immediately and rapidly.

1. Cut tissue into appropriate pieces using a single-edge razor blade if needed (*see* **Note 5**).
2. Freeze tissue directly onto cryostat chunks by embedding in OCT mounting medium, either on the cold shelf of the cryostat or over dry ice (*see* **Note 5**).
3. Cut cryostat sections (typically 10–30 μm) of tissue and thaw-mount onto gelatin-coated microscope slides (*see* **Note 6**). Appose tissue samples to a phosphor imaging screen (*see* **Subheading 3.5.**).

3.5. Phosphor Imaging

The methods below are based on the use of the Cyclone Phosphor Imager. For quantitative data, the tissue samples have to be coexposed with radioactive standards (*see* **Note 7**).

1. For quantification, prepare standards using the stock solution (*see* **Note 8**).
2. Mount microscope slides bearing tissue sections onto card, together with a microscope slide bearing the radioactive standards, and place in the exposure cassette.
3. Choose the size and sensitivity of the screen appropriate for the application (*see* **Tables 1** and **2** and **Note 3**).
4. Prepare the phosphor screen for imaging as indicated by the manufacturers. Erase the phosphor screen prior to exposure using the light box (*see* **Note 9**).
5. Place the phosphor screen against the radioactive samples in the exposure cassette (*see* **Note 10**). Close the cassette and leave for exposure for an appropriate length of time (*see* **Note 11**).
6. Remove the exposed screen from the cassette and place in the Cyclone scanning carousel. Load the carousel in the Cyclone Phosphor Imager (*see* **Notes 12** and **13**). Start OptiQuant image acquisition software.

3.6. Image Analysis

Analyze the autoradiograms using the OptiQuant image analysis software. To quantify data, regions of interest (ROIs) are drawn on the image. Different drawing tools are available in OptiQuant, such as a rectangle, an ellipse, or an irregularly-shaped drawing tool to measure amount of radioactivity in the whole or in discrete structures of the tissue section. Data will be expressed as Digital Light Unit (DLU) or DLU/mm^2. From the standard curve, the calibration factor DLU/fmol is obtained (*see* **Fig. 4.**).

1. Construct the standard curve by drawing regions around the spots using the ellipse-shaped tool. Draw a background region to enable background subtraction (*see* **Note 14**).

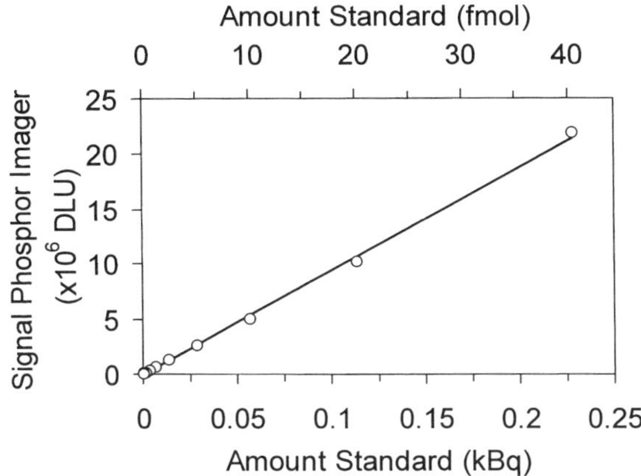

Fig. 4. Example of standard curve obtained using phosphor imaging. Known amounts of radioactivity from the radioligand stock were spotted (2 μL) onto polyethylene-backed absorbent paper and co-exposed with the tissue samples. From the known specific activity of the radioligand stock solution, the corresponding fmol radioligand could be calculated. The data have been fitted using linear regression ($r^2 = 0.998$).

2. Draw regions on the image for the tissue sections. Select an appropriate drawing tool for the shape of the whole tissue or the discrete region of the tissue. Draw a background region to enable background subtraction (*see* **Note 15**).

3.7. Data Analysis

Data from binding experiments are analyzed using nonlinear iterative curve fitting programs to calculate saturation parameters (affinity [K_D], density [B_{max}] and Hill slope [nH]) and association parameters (half-time for association [$t_{1/2}$] and the association rate constant [k_{obs}]). Nonlinear regression analysis is performed using the KELL software package (Biosoft, http://www.Biosoft.com) which contains the Equilibrium Binding Data Analysis (EBDA) and the LIGAND programs ***(10,11)***. The use of these programs for radioligand data analysis is described in Chapter 5.

4. Notes

1. The screens are re-usable and are said to have an indefinite lifetime, and can last for years if handled with care ***(3)***. However, it has been reported that some deterioration of the screens can take place with repeated use over time ***(12)***.
2. BSA is included, as this can minimize NSB using in vitro assays for transmitters such as peptides, but it may be omitted when using [^{18}F]-labeled drugs to avoid plasma binding, and hence loss of free radioligand.

3. Four different screen types are available for the Cyclone Phosphor Imager (**Table 1**). The MP, SR, and ST screens are suitable for applications involving PET radioligands. Owing to the high energy of the PET radionuclides, the difference in resolution of the different screens will probably not be significant. We have used the SR or the ST screen for our ^{18}F work, and both yield good images. For scanning, the screens are loaded in a scanning carousel and because the screens are available in different sizes, a separate scanning carousel for each screen size is needed.
4. Time to reach equilibrium has to be determined in association experiments (*see* **Subheading 3.3.**). However, when working with short-lived radionuclides, the rapid decay may limit the time available for incubation. This is most apparent when working with ^{11}C labeled radioligands, where incubation times of <1 h, i.e., less than three half-lives of the radionuclide, are typically used ***(7,13)***. On the other hand, with ^{18}F there is scope for longer incubation times, e.g., using the three half-lives limit as above would allow for incubation times for up to 5.5 h.
5. Tissue is cut and orientated on the chucks to give the desired cross section (transverse, coronal, or sagittal).
6. When cutting radioactive tissue, there is a potential risk for contamination of the cryostat, especially the cryostat blade. When working with short-lived PET radionuclides such as ^{11}C and ^{18}F, this will not be a significant problem because the major part of the radioactivity will decay within 24 h (^{11}C) and 48–72 h (^{18}F). However, if sectioning tissue labeled with long-lived radionuclides, contamination checks and decontamination procedures must be in place to minimize the risk of cross-contamination of the tissue.
7. Calibrated radioactive standards should be prepared from the same radionuclide as that used in the experiment so that the standard will have the same radioactive half-life and decay characteristics as the sample. Consequently, when working with short-lived PET radionuclides, calibrated standards must be made immediately prior to the exposure of the screen. This is best accomplished by using the radioligand stock solution. Furthermore, co-exposing tissue samples with the standards will ensure that the samples and standards have had the same exposure time (*see* **Note 11**) and the same level of potential signal fading (*see* **Note 12**).
8. Standards are prepared by serial dilution of the radioligand stock solution in assay buffer. The highest concentration is determined empirically depending on the experiment, but as a guide the standard curve in **Fig. 4.** was constructed by dilution (1:55) of the radioligand stock solution (6.3 MBq/mL) for the highest standard concentration, and subsequent concentrations were obtained by serial dilution (1:2). Aliquots (2 μL) of the standards were spotted onto polyethylene-backed absorbent paper (e.g., Benchkote or BenchGuard, VWR International Ltd., http://www.vwrcanlab.com), mounted on a microscope slide, and allowed to dry. From the known specific activity of the radioligand, the corresponding fmol of radioligand could be calculated.
9. Phosphor screens are sensitive to light. White light will erase them, and ultraviolet (UV) light will charge them. Store the screens at room temperature in a dry environment, away from all sources of light and β, γ and X-ray emissions. Always

erase the screen using a light box before use to ensure that the phosphor crystals are at the ground state.

10. Exposure starts immediately upon contact with the radioactive sample. Try to avoid adjusting the position of the screen after contact with the samples, because this may result in a blurred image.
11. The exposure time has to be estimated. The risk of saturating the screens is low because of the increased dynamic range compared with radiation-sensitive film. The optimum exposure time depends on the energy of the radionuclide and the amount of radioactivity in the sample, the sensitivity of the phosphor screen (**Table 1**), and the signal-to-background ratio needed to be able to obtain good quantitative data. When working with short-lived PET radionuclides, the fast decay introduces a limit to how long it is useful to expose the screens. With exposure times of three to four half-lives, 88–94% of the radioactive signal has been acquired by the phosphor screen, i.e., longer exposure time than 60–80 min and 5–7 h for ^{11}C and ^{18}F, respectively, will not increase the signal. Consequently, longer exposure times will only increase the background owing to background radiation and a loss of signal can occur as a result of the effect of fading (*see* **Note 12**). We have found that for ^{18}F, exposure times of 3–4 h are sufficient to yield good images.
12. The screens will lose the intensity of stored data with time (signal fading). For example, about 20% of the phosphor imager signal obtained from ^{14}C faded during the first 3 h *(2)*. Therefore, it is recommended that the screens be analyzed as soon as possible after finished exposure. When setting up a phosphor imaging protocol for PET radioligands, the effect of fading should be taken into account, e.g., exposing samples overnight could potentially mean that signal intensity can be lost (*see* **Note 11**).
13. When transferring the screen from the cassette to the scanning carousel, turn off any unnecessary ambient light. The screens are very sensitive to white light. Direct exposure to bright light can erase up to 90% of the stored image in less than 1 min.
14. For the standard curve, draw a region that will encompass the whole spot of the most concentrated standard. Then simply use the copy-and-paste option in the OptiQuant data analysis software for the rest of the standard curve concentrations. Subtracting the background region will give total activity in each ROI (*see* **Fig. 4**).
15. For total activity of the tissue section, the method used for the standard curve can be used to yield data as DLU/section (*see* **Note 14**). Use the standard curve to interpolate tissue data as fmol/section. However, data can also be expressed as DLU/mm^2, which is especially useful when, for example, analyzing binding to discrete structures of the tissue section. Using the standard curve data will consequently be expressed as fmol/mm^2.

Acknowledgments

Supported by grants from the British Heart Foundation, Royal Society, and MRC.

References

1. Davenport, A.P. (2000) Endothelin receptors. *IUPHAR Compendium of Receptor Characterisation and Classification*, 2nd Edition. IUPHAR Media, London, UK: 182–188.
2. Johnston, R.F., Pickett, S.C., and Barker, D.L. (1990) Autoradiography using storage phosphor technology. *Electrophoresis* **11**, 355–360.
3. Upham, L.V. and Englert, D.F. (1998) Radionuclide imaging. In *Handbook of Radioactivity Analysis* (L'Annunziata, M. F., ed.). Academic, San Diego, CA: pp. 647–692.
4. Chen, S.-F., Richards, H. K., Smielewski, P., et al. (2004) Relationship between flow-metabolism uncoupling and evolving axonal injury after experimental traumatic brain injury. *J. Cereb. Blood Flow Metab.,* **24,** 1025–1036.
5. Johnström, P., Harris, N. G., Fryer, T. D., et al. (2002) ^{18}F-Endothelin-1, a positron emission tomography (PET) radioligand for the endothelin receptor system: radiosynthesis and *in vivo* imaging using microPET. *Clin. Sci.* **103(Suppl 48)**, 4S–8S.
6. Johnström, P., Fryer, T. D., Richards, H. K., et al. (2004). Positron emission tomography using ^{18}F-labelled endothelin-1 reveals prevention of binding to cardiac receptors owing to tissue-specific clearance by ET_B receptors *in vivo*. *Br. J. Pharmacol.* **144,** 115–122.
7. Sihver, W., Sihver, S., Bergström, M., et al. (1997) Methodological aspects for *in vitro* characterisation of receptor binding using ^{11}C-labeled receptor ligands: a detailed study with the benzodiazepine receptor antagonist [^{11}C]Ro 15-1788. *Nucl. Med. Biol.* **24**, 723–731.
8. Peter, M. G. and Davenport, A. P. (1996) Characterization of the endothelin receptor selective agonist, BQ3020 and antagonists BQ123, FR139317, BQ788, 50235, Ro462005 and bosentan in the heart. *Br. J. Pharmacol.* **117**, 455–462.
9. Karet, F. E., Kuc, R. E. and Davenport, A. P. (1993) Novel ligands BQ123 and BQ3020 characterize endothelin receptor subtypes ET_A and ET_B in human tissue. *Kidney Int.* **44**, 36–42.
10. Munson, P. J. and Rodbard, D. (1980) Ligand: a versatile computerized approach for characterization of ligand-binding systems. *Anal. Biochem.* **107**, 220–239.
11. McPherson, G. A. (1985) Analysis of radioligand binding experiments. A collection of computer programs for the IBM PC. *J. Pharmacol. Methods* **14**, 213–228.
12. Reinprecht, I., Windisch, M., and Fahnestock, M. (2002) Deterioration of storage phosphor screens with use. *J. Label. Compd. Radiopharm.* **45**, 339–345.
13. Sihver, S., Sihver, W., Bergström, M., et al. (1999) Quantitative autoradiography with short-lived positron emission tomography tracers: a study on muscarinic acetylcholine receptors with *N*-[^{11}C]-4-piperidylbenzilate. *J. Pharmacol. Exp. Ther.* **290**, 917–922.

11

Dynamic In Vivo Imaging of Receptors in Small Animals Using Positron Emission Tomography

Peter Johnström, Tim D. Fryer, Hugh K. Richards, Olivier Barret, and Anthony P. Davenport

1. Introduction

1.1. Positron Emission Tomography

Positron emission tomography (PET) is a functional imaging technique that is used to study biological processes in vivo. Data obtained in a PET scan can provide information regarding tissue physiology or pathophysiology, as well as pharmacokinetic and pharmacodynamic information. It is the most sensitive technique available to image and quantify receptor distributions in vivo, and it has been used extensively to study major neurotransmitter systems such as the dopamine, serotonin, benzodiazepine, opiate, and cholinergic systems (*1*). Over the years, PET has increasingly been recognized as a very powerful tool to accelerate development and assessment of existing and novel drugs (*2–7*).

To utilize PET compounds labeled with short-lived positron-emitting radionuclides (*see* **Table 1**) have to be synthesized prior to administration of the radioligand into the healthy subject or patient. The subsequent biodistribution of the radioligand in the body is imaged by the PET scanner. This imaging device detects, in coincidence, the two co-linear γ-photons formed in the annihilation of the emitted positron with an electron (*see* **Fig. 1**). Data acquired in a PET scan are typically reconstructed into a series of images describing the spatiotemporal distribution of radioactivity in the body (*see* **Figs. 2** and **3**). Using image analysis software such as Analyze (*8*) (http://www.mayo.edu/bir/

From: *Methods in Molecular Biology, vol. 306: Receptor Binding Techniques: Second Edition*
Edited by: A. P. Davenport © Humana Press Inc., Totowa, NJ

Table 1
Examples of Positron-Emitting Radionuclides

Radionuclide	Half-life (min)	Positron energy (MeV)[a]	Cyclotron produced starting materials[b]
^{11}C	20.4	0.96	$^{11}CO_2$, $^{11}CH_4$, $^{11}CN^-$
^{13}N	9.96	1.19	$^{13}NH_3$, $^{13}NO_x$
^{15}O	2.07	1.72	$^{15}O_2$
^{18}F	109.8	0.635	$^{18}F^-$, $^{18}F_2$
^{64}Cu	12.7 h	0.653	$^{64}Cu^+$, $^{64}Cu^{2+}$

The half-life of the radionuclide selected for labeling should be compatible with the rate of the biological process to be investigated. For imaging of receptors, radioligands are commonly labeled with ^{11}C and ^{18}F. For a comprehensive listing of radioligands developed for position emission tomography (PET), *see* **ref. *26***, and for an introduction to radiolabeling methods, *see* chapters in **ref. *27***.

[a]The higher the energy, the longer the distance the positron will travel before undergoing an annihilation process with an electron, i.e., the poorer the image resolution (*see* **Fig. 1**).

[b]Examples of the relatively simple starting materials available for radiosynthesis of PET radioligands. The challenge for the radiochemist is to be able to synthesize these radioligands in sufficient yield and with a high specific activity.

Software/Analyze/Analyze.html), time-radioactivity curves describing the uptake kinetics of the radioligand in regions of interest (ROIs) are constructed (*see* **Fig. 4**). These curves provide information on uptake, clearance (if any), and the concentration of radioligand in various tissues as a function of time. Furthermore, from mathematical modeling or graphical analysis of the data, parameters such as rate constants for plasma-to-tissue influx (K_1), tissue-to-plasma efflux (k_2) and binding to the receptor (k_3 and k_4), as well as the binding potential (B_{max}/K_D) and volume of distribution, can be derived ***(9–11)***.

1.2. Small-Animal Imaging

Historically, PET scanners were designed for human use with a typical resolution of the order of 5 mm (***12***). This limitation in spatial resolution has made it difficult to image small animals such as rats and mice. However, the potential of small-animal imaging to study the biodistribution and kinetics of radioligands has been recognized for some time (*see* **ref. *13***), stimulating the research and development of PET imaging systems for small animals. A crucial part of this development has been the advances in detector technology which now permit the design and construction of dedicated scanners for small animals with a spatial resolution of 1–2 mm ***(14,15)***. A number of these systems are now commercially available, such as the microPET (*see* **Fig. 5**; Concorde MicroSystems, http://www.Concorde.ctimi.com) the Quad-HIDAC

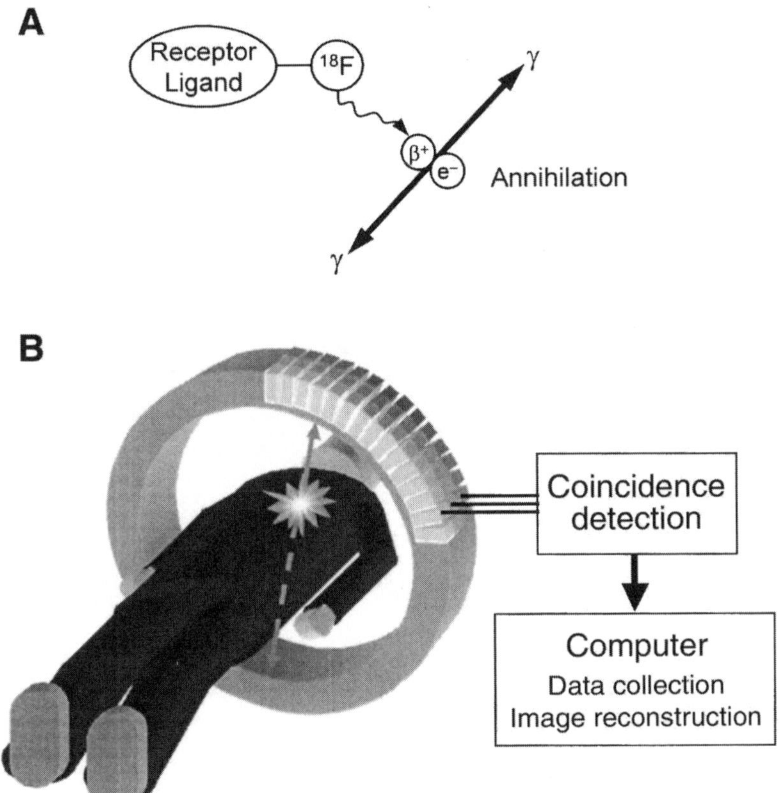

Fig. 1. Principle of positron emission tomography (PET). (**A**) When the radionuclide decays, a positron (β^+) is emitted which will travel a short distance in the tissue before annihilating with an electron (e^-). The annihilation yields two opposing γ-photons (511 keV) that can escape from the body and be detected externally. (**B**) The PET scanner consists of rings of detectors that detect the γ-photons (each photon is referred to as a single). For the annihilation to be counted as a true event, the two singles have to be registered in coincidence, i.e., within a short time interval (typically 6–12 ns). This event is stored in a computer. At end of scanning, the stored data are reconstructed into images describing the spatiotemporal distribution of radioligand in the body (*see* **Figs. 2** and **3**).

(Oxford Positron Systems, http://www.oxpos.co.uk), the YAP-(S)PET (ISE, http://www.ise-srl.com) and the Mosaic (Philips Medical Systems, http://www.medical.philips.com).

The improvement in spatial resolution obtained with these systems compared with human scanners allows the delineation of discrete organs and their larger substructures within rats and mice (*see* **Figs. 2** and **3**) and, consequently, pro-

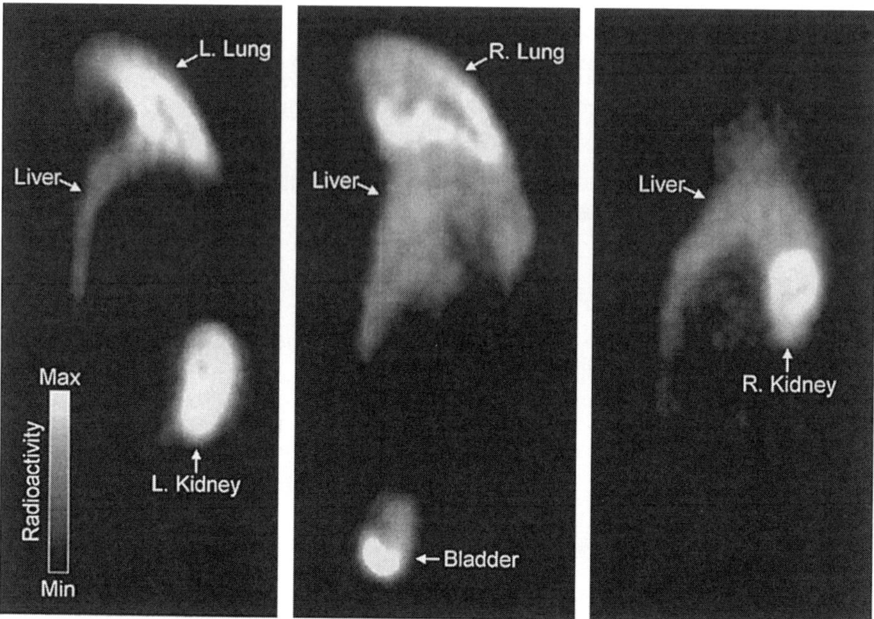

Fig. 2. Images showing the biodistribution of [^{18}F]-ET-1 (***28***), a radioligand for the endothelin (ET) receptor, in rat in three different sagittal planes. High levels of uptake could be visualized in organs with high densities of ET receptors, such as the lung and kidney. The experiment was performed using an anesthetised Sprague–Dawley rat and the microPET P4 scanner.

vides the means to study normal and animal models of disease using PET. Furthermore, it will be possible to perform longitudinal studies in these models to monitor disease progression or effect of treatment in the same animal. Small-animal imaging enables testing and validation of novel PET radioligands in vivo, and the biodistribution obtained can be directly correlated with the anatomical distribution of the radioligand in ex vivo and in vitro autoradiography (*see* Chapter 10). It will be possible to acquire complete pharmacokinetic data in a single animal, i.e., using small-animal imaging will significantly reduce the number of animals needed for these types of experiments as well as reduce the effect of interanimal variation. The biodistribution data may also provide information for a dosimetry assessment of the radioligand, such as major organs of uptake and residence times.

1.3. Quantification of Imaging Data

PET has the potential to produce quantitative data. In order to do so, various corrections need to be applied to the image data and an appropriate input func-

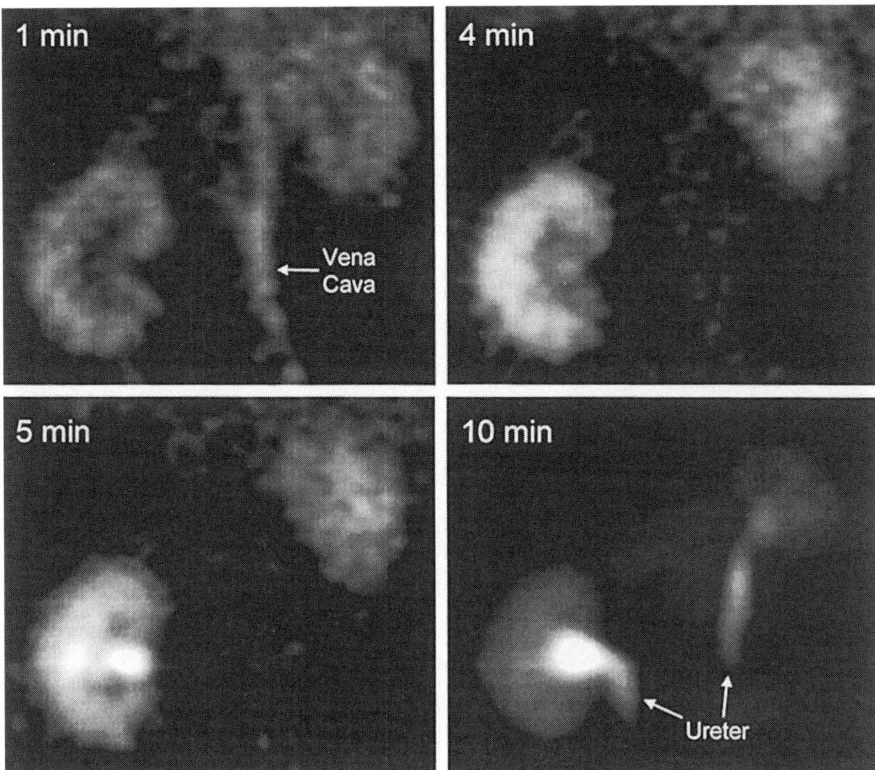

Fig. 3. Images showing the spatiotemporal distribution of the metabolic tracer [^{18}F]fluorodeoxyglucose ([^{18}F]FDG) in rat kidney. Uptake in suborgan structures in the kidney could be visualized with the microPET. The dynamic data are consistent with a rapid clearance of radioactivity from the circulation by renal excretion. The experiment was performed using an anaesthetised Sprague–Dawley rat and the microPET P4 scanner.

tion or reference tissue must be defined. The following corrections need to be applied to the image data to make the images as true a representation of the tracer distribution as possible:

1. Background. This is especially relevant to scanners with lutetium oxyorthosilicate (LSO) detectors, such as microPET, as this detector material is naturally radioactive.
2. Dead time. Owing to the finite time it takes to process an event (*see* **Fig. 1**), the response of any PET system is not linear with increasing activity. This needs to be corrected for, using either a direct method or a model based on singles and/or coincidence rates.

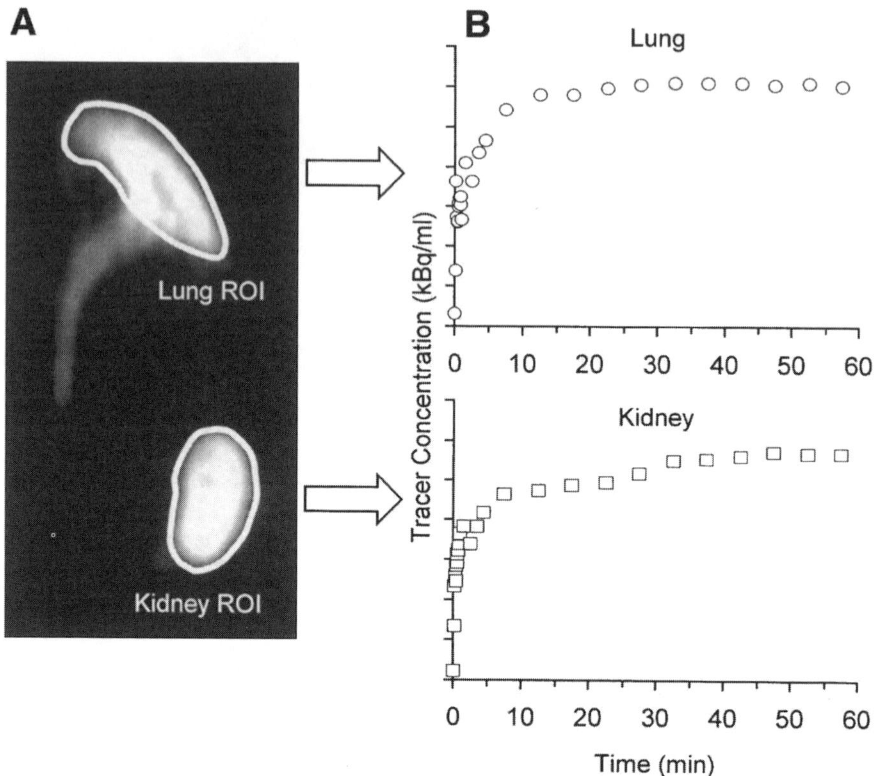

Fig. 4. Time-radioactivity curves describing the uptake kinetics of the radioligand in regions of interest (ROIs) are constructed from the dynamic data using image analysis software (e.g., Analyze). (**A**) A ROI is drawn, encompassing the area of uptake in all image planes containing the organ, thereby creating a ROI volume. The concentration of radioactivity in the ROI volume for each time frame can then be calculated by the software. (**B**) Using the results from the image analysis, graphs describing the concentration of radioactivity as a function of time in each ROI can be constructed.

3. Randoms. These are coincidence events (*see* **Fig. 1**) formed from photons emanating from separate annihilations. They do not provide accurate spatial information and are not proportional to activity. Correction is achieved by acquiring a pure randoms data set using a delayed coincidence window and subtracting these events from the main dataset.
4. Normalization. Owing to variability in individual detector response and the geometry of the scanner, the sensitivity along the lines of response passing through the object is not uniform. Using data acquired from sources of known position and geometry, relative sensitivity factors can be calculated for all the lines of response. These are applied to the data in the image reconstruction stage.

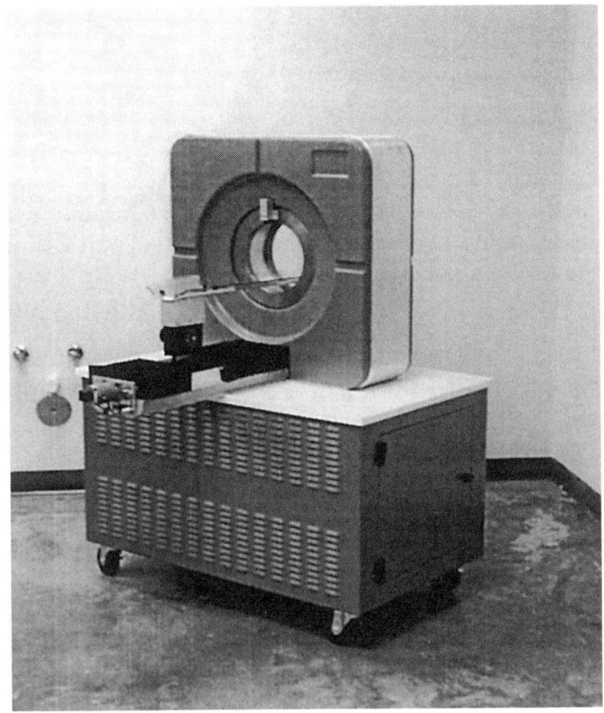

Fig. 5. The microPET P4 scanner from Concorde MicroSystems (http://www.Concorde.ctimi.com) is a four-ring system with a total of 10,752 detectors. The system operates in three-dimensional mode and data is acquired in list mode. The scanner port diameter is 22 cm, and the axial field of view is 7.8 cm. The animal bed is computer controlled, which facilitates whole-body scanning. A spatial resolution of 1.8 mm at the center of the field of view has been reported for this system (***24***).

5. Scatter. Compton scattering of the annihilation photons occurs in the materials located both inside and outside the field of view. The main component of this occurs in the animal, but there will also be scattering in, for example, the bed. These events produce incorrect spatial information. Using a narrow energy window preferentially discriminates scatter events from true events, as the scattered photons have lower energy. However, there is always some residual scatter that has to be corrected for using software. There are various methods available, but the most popular is the single scatter modeling.
6. Attenuation. The aforementioned Compton scattering results in not only the detection of scattered events but also in the loss of events from their original trajectory. This is a spatially variant phenomenon that results in a suppression of the signal in the areas that have the longest attenuation path lengths going through them, typically the center of the object. Attenuation is corrected for by acquiring

a transmission scan with either a rotating radioactive source or with a computed tomography (CT) scanner. In both cases, the attenuation along a line of response is calculated and this is applied as a multiplicative correction to the data.
7. Sensitivity. Even after normalization, the sensitivity of the planes within the image is not uniform. Using a uniform phantom with a known activity concentration, slice sensitivity factors and a global factor to convert the reconstruction image into radioactivity concentration (kBq/mL) can be determined.
8. Decay. To obtain the tracer concentration as a function of time during the scan, as opposed to the level of radioactivity, a multiplicative correction needs to be applied to the kinetic data for the decay of the positron-emitting isotope.

If kinetic modeling is to be applied to the data, then invariably an input function or a reference tissue time-activity curve needs to be determined. The choice of the two depends on the model being used (*see* **refs.** ***9–11***). The input function is an estimate of the tracer delivered to the tissue in blood. This can be obtained through direct blood sampling or by using an image-based method with a region located over a vascular region, e.g., blood pool of the heart. Reference tissue analysis is often applied in neurological studies. The reference region is assumed to have similar kinetic properties to the ROI except that the binding or metabolic reaction of interest is absent in the reference region.

1.4. Practical Aspects of Small-Animal Imaging

Animals have to be immobilized during scanning, hence some form of anesthesia has to be administered. The type and level of anesthetics can potentially affect animal physiology, and consequently might affect the uptake kinetics of the radioligand (***16,17***). Furthermore, although the scanned animal is anesthetized, motion artefacts are still possible as a result of breathing or gut motility. For brain scans in rodents, the head is commonly held in a purpose-built, plastic stereotaxic frame incorporating ear bars and a bite bar. Organ movements (e.g., beating heart) can potentially contribute to loss in resolution, i.e., making the reconstructed image blurred. To minimize the effect of the moving heart on image resolution, techniques such as list-mode gated PET scanning can be used (***18***).

Animal hemodynamics can be monitored on-line using systems such as the MP100WSW (BioPac System Inc., http://biopac.com). We have successfully used the MP100WSW system in conjunction with microPET scanning to monitor BP and heart rate in the rat. The blood pressure (BP) transducer was connected to a canulated femoral artery for BP monitoring and using the Acq*Knowledge* III acquisition software (BioPac System Inc.) the heart rate could be calculated from the BP data and concurrently be displayed on-line.

Radioligands are injected intravenously, with the site for injection located preferably outside the field of view of the scanner. In rats, femoral or tail veins are commonly used, whereas for mice, injection is performed into a warmed tail vein. In the New Zealand white rabbit, the marginal ear vein may be used. In humans, radioligands are typically injected as a bolus followed by a saline flush to ensure that the whole dose of radioligand has been administered. However, because of the rapid circulation time in small animals, the short time delay between injection and saline flush could potentially give rise to a double peak at the beginning of the time-activity tissue curve. This double peak will make it difficult to perform an accurate quantitative analysis of the data in tracer kinetic modeling. For this type of experiment, injection of radioligand without saline flush may be a better option.

As mentioned previously, kinetic modeling of the data requires the determination of the input function, i.e., the concentration of unmetabolized radioligand in plasma. This requires blood sampling and analysis of metabolites. Implementing this into a small-animal imaging protocol is very complex and labor-intensive. First, as a result of the rapid circulation in small animals, it is necessary to perform very rapid initial sampling so that the peak of radioactivity in the blood can be well defined. Second, the total blood volume available in small animals limits the number of samples that can be collected. And finally, methods for rapid analysis of metabolites have to be developed using the small sample volumes available. In spite of these technical challenges, manual blood sampling has been reported in small-animal PET imaging experiments using [^{18}F]-fluorodeoxyglucose (FDG) (***19***) and [^{11}C]flumazenil, including analysis of metabolites (***20***). A number of systems for on-line blood sampling on small animals have been developed to simplify determination of the input function (*see* **refs. *13,21***). We have evaluated the potential of the commercially available BetaMicroProbe (Biospace, http://www.biospace.fr) to monitor radioactivity in blood. This probe utilizes a scintillating fiber for the detection of the positron emitted from PET radioligands, and its potential to follow the dynamic distribution of radioactivity in tissue has been demonstrated (***22***). For monitoring of blood radioactivity, the right femoral artery of an anesthetised rat was cannulated for conventional blood sampling, and the BetaMicroProbe was carefully inserted into the left femoral artery and advanced such that the tip was positioned high in the abdominal aorta. Discrete blood sampling from the right artery was used for comparison and quantification of MicroBetaProbe data. A representative blood curve is shown in **Fig. 6** and, as can be seen, a well-defined blood peak was obtained. The probe is very light-sensitive, and therefore has to be protected from ambient light in the room to minimize background and noise levels. For this purpose, the equipment is provided with a piece of black fabric to cover the animal around the area of the probe. However, we also had to dim the light significantly in the room to be able to obtain useful data. In addition,

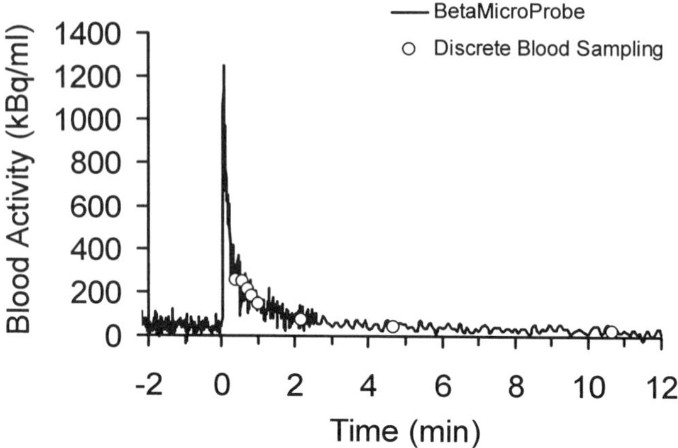

Fig. 6. Blood curve for a bolus injection of 10.5 MBq [^{18}F]-ET-1 (*28*) in an anesthetised Sprague–Dawley rat obtained using the BetaMicroProbe. A 250-μm probe was carefully inserted into the left femoral artery and advanced such that the tip was positioned high in the abdominal aorta. Data was collected using a sampling frequency of 1 s. Discrete blood samples were collected from the right femoral artery and measured in a well counter. For quantification, MicroProbe data (cps) was normalized to the discrete blood sampling data (kBq/mL).

discrete blood sampling still has to be performed in a separate arterial line to enable quantification of BetaMicroProbe data.

Obviously, obtaining the input function for mice will be even more difficult, and hence the use of imaged-based methods has been proposed (*23*).

2. Materials

1. Animal surgery laboratory equipped for handling PET isotopes.
2. Heating blanket with temperature control (e.g., a homeothermic blanket system from Harvard Apparatus, http://www.harvardapparatus.com) (*see* **Note 1**).
3. Equipment for administration of anesthesia (*see* **Note 2**).
4. Hemodynamic monitoring equipment if needed (e.g., MP100WSW with Acq*Knowledge* III acquisition software)
5. Small-animal PET scanner (e.g., microPET P4 scanner, Concorde MicroSystems, http://www.Concorde.ctimi.com) (*see* **Fig. 5**).
6. Radiolabeled PET ligands.
7. Receptor antagonists for pharmacological intervention, i.e., preblocking or displacement of radioligand binding, if needed.
8. Dose calibrator for measuring injected dose (e.g., Capintec Dose Calibrators, http://www.capintec.com).
9. Well counter for measuring radioactivity in tissue and blood samples if needed. Preweighed test tubes to collect tissue and blood.

10. Equipment for ex vivo autoradiography of tissue sections if needed (*see* Chapter 10).

3. Methods

All animal experiments must be conducted in accordance with the appropriate legislation for experimental work on animals and comply with guidelines of the local animal ethics committee. In the United Kingdom, experimental work on animals is regulated by the UK Animal Scientific Procedures Act, 1986.

Here we present a method for the imaging of rats in the microPET P4 scanner *(24)* (*see* **Fig. 5**).

3.1. Animal Preparation

1. Anesthetise the rat (*see* **Note 2**).
2. Place rat on heating blanket and insert rectal thermometer.
3. Perform surgery, i.e., cannulate a femoral vein using 0.4-mm internal diameter polyethylene tubing (SIMS Portex, UK) for administration of radioligand. Insert injection needle in tubing to allow for connection with syringe (*see* **Note 3**).
4. If blood sampling and/or BP monitoring is going to be performed, cannulate a femoral artery for each procedure using 0.4-mm internal diameter polyethylene tubing.

3.2. MicroPET Scanner Preparation

1. Set up the microPET for scanning. Perform a quality assurance scan with a low-activity, rotating ^{68}Ge source to check the sensitivity of the scanner. Compare with previous readings.
2. Acquire a blank scan. This will be used in combination with the transmission scan to calculate attenuation correction factors.
3. Set up acquisition protocol for the emission study (*see* **Note 4**).

3.3. Small-Animal Imaging

1. Transfer the animal to the scanner bed and fix the animal using an appropriate method. Position the animal in the microPET scanner using the computer controlled bed so that the organs of interest are encompassed by the field of view.
2. If BP monitoring is going to be performed, connect the BP probe to the arterial line and set up an acquisition method as specified by the system manual.
3. Start BP monitoring to obtain baseline data.
4. Perform transmission scan using rotating ^{68}Ge point source (*see* **Note 5**).
5. Draw up an appropriate amount of radioligand from the delivered stock solution. Measure the activity in the syringe using the dose calibrator (*see* **Note 6**).
6. Start scanner acquisition and inject radioligand as a bolus and flush with saline if appropriate (*see* **Subheading 1.4** and **Notes 7** and **8**).
7. Perform blood sampling if included in the experimental protocol (*see* **Subheading 1.4**)

8. At end of scanning, remove the animal from the scanner bed and fixation system (*see* **Note 9**).

3.4. Image Reconstruction

1. Histogram the list-mode data into sinograms.
2. Apply aforementioned data corrections (*see* **Subheading 1.3**). Some of these will be integrated in either the histogramming or image reconstruction software.
3. Reconstruct the data using an appropriate reconstruction algorithm (*see* **Note 10**).

3.5. Image Analysis

To construct time-activity curves images are analyzed in Analyze (***8***).

1. Transfer image data to Analyze format.
2. Construct a multi-time frame image volume from the dynamic images.
3. Select which time frame or summed time frames to draw the ROI in (*see* **Note 11**).
4. Draw a ROI in all planes encompassing the area of uptake in one orientation (transverse, sagittal, or coronal). Check ROI in the other orientations not used for drawing and correct size if necessary (*see* **Fig. 4**).
5. Calculate the concentration of radioactivity in the ROI volume for each time frame using the multi-time frame image volume.
6. Construct time-activity graphs (*see* **Fig. 4**).
7. Perform kinetic modeling of data if required (*see* **refs. *9–11***).

4. Notes

1. The body temperature of an anesthetized animal has to be controlled and maintained within the normal range throughout the experiment. This is usually performed using a heated blanket, with temperature control provided by a rectal thermometer.
2. We have used isofluorane to anesthetise rats, mice, and rabbits. Typically, rats were anesthetised with 3% isofluorane vaporised in N_2O/O_2 (0.8/0.4 L/min) and maintained with 2% isofluorane during surgery. During PET scanning, anesthesia was reduced to 1.5%–2% isofluorane in N_2O/O_2 (0.8/0.4 L/min). The choice of anesthetic is determined primarily by the nature of the experimental procedure, but in general gaseous or intravenous agents are preferred. If supplemental dosing is required over the course of scanning, which may be several hours, intramuscular or intraperitoneal injections may cause movement artefacts.
3. If the experimental protocol includes postadministration of unlabeled antagonists, i.e., displacement of radioligand binding, this can be performed using the same line as that for radioligand injection. However, there is a potential risk of residual radioactivity still present in the line being injected, especially if the line had not been flushed with saline after radioligand administration (*see* **Subheading 1.4**). This may potentially complicate the interpretation of radioligand displacement

kinetics. For these experimental protocols, a second venous line for administration of the unlabeled antagonist may be a better option.

4. For the microPET, data is acquired in list mode. Usually an acquisition protocol consists of a number of time frames. These should be set up to best reflect the biological process studied. Typically, short time frames are used at the start of the experiment to follow fast uptake dynamics, and longer time frames are used at the end of the experiment when slow or no changes in levels of tissue uptake are expected. If a pharmacological intervention is going to be performed, e.g., displacement of binding, data should be collected using short time frames to allow for the detection of rapid changes in the level of radioactivity in the tissues. However, for the microPET, because data is collected in list mode, data can be reframed if needed, enabling the re-analysis of time-activity curves.
5. To be able to get quantitative data, a number of corrections have to be applied to the collected data (*see* **Subheading 1.3**). To be able to correct for attenuation and scatter, a transmission scan has to be performed, preferably prior to the injection of the radioligand. The ^{68}Ge point source is rotated around the animal and transmission data is collected.
6. The injected dose is determined by measuring the syringe before and after injection in the dose calibrator.
7. If the effects of a pharmacological intervention, such as preblocking or displacement of binding, are going to be studied, the unlabeled competitor should be infused prior to or at an appropriate time after injection of the radioligand, respectively (*see* **Note 3**).
8. Time for scanning is obviously determined by the time needed to follow the biological process studied. However, the half-life of the radionuclide used for labeling the radioligand limits the time available for scanning. Typically, scanning times up to 3–4 half-lives of the radionuclide are feasible, i.e., 60–80 min for ^{11}C and 5–7 h for ^{18}F.
9. The treatment of the animal after scanning has finished is dependent on the experimental protocol. If the animal is part of a longitudinal study, it should be left to recover according to an approved protocol and then kept for subsequent scanning. Alternatively, if organs are going to be removed for ex vivo analysis, the animal is euthanized using an approved method and dissected. Tissues are weighed and assayed for the amount of radioactivity using a well counter. For autoradiography of tissue sections, *see* Chapter 10.
10. Various reconstruction algorithms are possible: Fourier rebinning (FORE) + 2D filtered backprojection (FBP) or 2D OSEM; 3D FBP; 3D OSEM; 3D MAP. These differ in the compromise between image resolution, noise, bias, and computational expense. We have used the 3D FBP algorithm (***25***), adapted in-house to work with data from the microPET P4 scanner.
11. In the Analyze program there are several options for image visualization. An average image for several time frames can be created using the Image Algebra option. This image can, in some cases, be a better option for drawing of the ROI, because averaging will reduce the noise level.

Acknowledgments

Supported by grants from the British Heart Foundation, Royal Society, and MRC. The University of Cambridge microPET system was acquired from a JREI grant from HEFCE and Merck Sharp and Dohme, Ltd.

References

1. Fowler, J. S., Ding, Y. S., and Volkow, N. D. (2003) Radiotracers for positron emission tomography imaging. *Semin. Nucl. Med.* **33,** 14–27.
2. Farde, L. (1996) The advantage of using positron emission tomography in drug research. *Trends Neurosci.* **19,** 211–214.
3. Fowler, J. S., Volkow, N. D., Wang, G. J., Ding, Y. S., and Dewey, S. L. (1999) PET and drug research and development. *J. Nucl. Med.* **40,** 1154–1163.
4. Burns, H. D., Hamill, T. G., Eng, W. S., Francis, B., Fioravanti, C., and Gibson, R. E. (1999) Positron emission tomography neuroreceptor imaging as a tool in drug discovery, research and development. *Curr. Opin. Chem. Biol.* **3,** 388–394.
5. Aboagye, E. O., Price, P. M., and Jones, T. (2001) *In vivo* pharmacokinetics and pharmacodynamics in drug development using positron-emission tomography. *Drug Discov. Today.* **6,** 293–302.
6. Eckelman, W. C. (2002) Accelerating drug discovery and development through in vivo imaging. *Nucl. Med. Biol.* **29,** 777–782.
7. Passchier, J., Gee, A., Willemsen, A., Vaalburg, W., and van Waarde, A. (2002) Measuring drug-related receptor occupancy with positron emission tomography. *Methods.* **27,** 278–286.
8. Robb, R. A., Hanson, D. P., Karwoski, R. A., Larson, A. G., Workman, E. L., and Stacy, M. C. (1989) Analyze: a comprehensive, operator-interactive software package for multidimensional medical image display and analysis. *Comput. Med. Imaging Graph.* **13,** 433–454.
9. Logan, J. (2000) Graphical analysis of PET data applied to reversible and irreversible tracers. *Nucl. Med. Biol.* **27,** 661–670.
10. Slifstein, M. and Laruelle, M. (2001) Models and methods for derivation of *in vivo* neuroreceptor parameters with PET and SPECT reversible radiotracers. *Nucl. Med. Biol.* **28,** 595–608.
11. Lammertsma, A. A. (2002) Radioligand studies: imaging and quantitative analysis. *Eur. Neuropsychopharmacol.* **12,** 513–516.
12. Zanzonico, P. (2004) Positron emission tomography: a review of basic principles, scanner design and performance, and current systems. *Semin. Nucl. Med.* **34,** 87–111.
13. Ingvar, M., Eriksson, L., Rogers, G. A., Stone-Elander, S., and Widén, L. (1991) Rapid feasibility studies of tracers for positron emission tomography: high-resolution PET in small animals with kinetic analysis. *J. Cereb. Blood Flow Metab.* **11,** 926–931.
14. Chatziioannou, A. F. (2002) Molecular imaging of small animals with dedicated PET tomographs. *Eur. J. Nucl. Med. Mol. Imaging.* **29,** 98–114.

15. Lewis, J. S., Achilefu, S., Garbow, J. R., Laforest, R., and Welch, M. J. (2002) Small animal imaging: current technology and perspectives for oncological imaging. *Eur. J. Cancer.* **38,** 2173–2188.
16. Matsumura, A., Mizokawa, S., Tanaka, M., et al. (2003) Assessment of microPET performance in analyzing the rat brain under different types of anesthesia: comparison between quantitative data obtained with microPET and ex vivo autoradiography. *Neuroimage* **20,** 2040–2050.
17. Toyama, H., Ichise, M., Liow, J. S., et al. (2004) Evaluation of anesthesia effects on [^{18}F]FDG uptake in mouse brain and heart using small animal PET. *Nucl. Med. Biol.* **31,** 251–256.
18. Croteau, E., Benard, F., Cadorette, J., et al. (2003) Quantitative gated PET for the assessment of left ventricular function in small animals. *J. Nucl. Med.* **44,** 1655–1661.
19. Shimoji, K., Ravasi, L., Schmidt, K., et al. (2004) Measurement of cerebral glucose metabolic rates in the anesthetized rat by dynamic scanning with ^{18}F-FDG, the ATLAS small animal PET scanner, and arterial blood sampling. *J. Nucl. Med.* **45,** 665–672.
20. Fryer, T. D., Beech, J. S., Hughes, J. L., et al. (2003) Imaging benzodiazepine receptors in control and stroked rat brains using [^{11}C]flumazenil and microPET. *Mol. Imaging Biol.* **5,** 107–108.
21. Lapointe, D., Cadorette, J., Rodrigue, S., Rouleau, D., and Lecomte, R. (1998) A microvolumetric blood counter/sampler for metabolic PET studies in small animals. *IEEE Trans. Nucl. Sci.* **45,** 2195–2199.
22. Zimmer, L., Hassoun, W., Pain, F., et al. (2002) SIC, an intracerebral β^+-range-sensitive probe for radiopharmacology investigations in small laboratory animals: binding studies with ^{11}C-raclopride. *J. Nucl. Med.* **43,** 227–233.
23. Huang, S. C., Wu, H. M., Shoghi-Jadid, K., et al. (2004) Investigation of a new input function validation approach for dynamic mouse microPET studies. *Mol. Imaging Biol.* **6,** 34–46.
24. Tai, Y. C., Chatziioannou, A., Siegel, S., et al. (2001) Performance evaluation of the microPET P4: a PET system dedicated to animal imaging. *Phys. Med. Biol.* **46,** 1845–1862.
25. Kinahan, P. E. and Rogers, J. G. (1989) Analytic 3D image-reconstruction using all detected events. *IEEE Trans. Nucl. Sci.* **36,** 964–968.
26. Iwata, R. (2004) Reference book 2004 © for PET radiopharmaceuticals. http://Kakuyaku.cyric.tohoku.ac.jp/indexe.html.Date accessed: December 9, 2004.
27. Welch, M. J. and Redvanly, C. S. (Eds.) (2003) *Handbook of radiopharmaceuticals—radiochemistry and applications*. John Wiley & Sons Ltd, Chichester.
28. Johnström, P., Harris, N. G., Fryer, T. D., et al. (2002) ^{18}F-Endothelin-1, a positron emission tomography (PET) radioligand for the endothelin receptor system: radiosynthesis and *in vivo* imaging using microPET. *Clin. Sci.* **103(Suppl 48),** 4S–8S.

Index

U

V

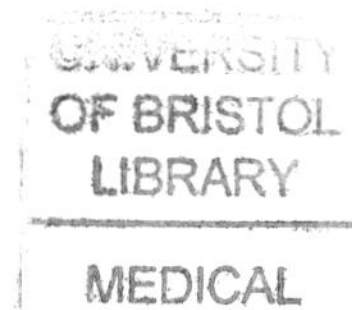
UNIVERSITY OF BRISTOL LIBRARY MEDICAL